CW00505166

ELECTRICAL AND ELECTRONIC PRINCIPLES II
SECOND EDITION

ANTHONY NICOLAIDES

B.Sc. (Eng.), C. Eng. M.I.E.E.
SENIOR LECTURER

P.A.S.S. PUBLICATIONS

PRIVATE ACADEMIC & SCIENTIFIC STUDIES LTD

Electrical and Electronic Principles II

First Published in Great Britain 1991 by
Private Academic & Scientific Studies Limited

ISBN 1 872684 34 3
SECOND EDITION 1996

Published also 1996
Electrical and Electronic Principles III

Printed and bound in Great Britain by
Hartnolls Limited, Bodmin, Cornwall

PREFACE

This book covers the topic of Electrical and Electronic Principles II.

This books is designed to assist the student wishing to master the subject of Electrical and Electronic Principles II. The series is easy to follow with minimum help. It can be easily adopted by a student who wishes to study it in the comforts of his home at his pace without having to attend classes formally; it is ideal for the working person who wishes to enhance his knowledge and qualification. Electrical and Electronic Principles II book, is divided into two parts. In Part I, the theory is comprehensively dealt with, together with many worked examples and exercises. A step by step approach is adopted in all the worked examples. Part II of the book, a special and unique feature acts as a problem solver for all the exercises set at the end of each chapter in Part I.

This book develops the basic concepts and skills that are essential for the Electrical and Electronic Principles II.

I am grateful to Mr. Myat Thaw Kaung, an excellent ex-student of mine, who typeset the manuscript superbly with great care on a desktop publishing system.

I am also grateful to Mr. Alex Yau for checking thoroughly this book.

I am also grateful to Mr. Robert Flegg for drawing the diagrams in this book.

This book covers the GNVQ syllabus.

Revised and improved edition

A. Nicolaides

ELECTRICAL AND ELECTRONIC PRINCIPLES II

PART I

CONTENTS

9. SOLUTIONS

MISCELLANEOUS

INDEX

THEORY PART I

ELECTRICAL AND ELECTRONIC PRINCIPLES II
UNIT NUMBER U86/329
BTEC

CIRCUIT THEORY

1. Applies circuit theory to the solution of simple circuit problems.
   a. Applies ohm's law to the solution of problems relating to series-parallel combinations of resistors.

THE UNIT OF CURRENT

The unit of current is the ampere, abbreviated by the symbol A. The ampere may be defined in terms of the force between conductors.

An ampere is defined as the constant current that flows in each of two infinitely-long parallel straight conductors of negligible cross-sectional area separated by a distance of one metre in *vacuo* when the force between these conductors is equal to 2×10^{-7} newtons per metre length. This definition is simply illustrated in Fig. 1.

1 A

1 m

1 A

Fig. 1   Two infinitely long conductors placed 1 m apart.

If the currents flow in the same direction, the force is attractive, if the currents flow in the opposite direction as shown in Fig. 1, the force is repulsive.

CURRENT FLOWS DUE TO THE EXISTENCE OF A POTENTIAL DIFFERENCE (VOLTAGE) BETWEEN TWO POINTS IN AN ELECTRICAL CONDUCTOR.

Electrical current is the flow of electrical charge and current flows through a resistor when a source of energy is supplied.

Let us consider a current where a d.c. source of e.m.f. E and negligible internal resistance is connected across a resistor R, as shown in Fig. 2.

Fig. 2 Conventional current and voltage.

Conventionally, it is assumed that the current I flows from the positive terminal of the e.m.f. source to the negative terminal as shown in Fig. 2.

It is observed that the current I flows from A to B via the resistor, therefore the point A is positive with respect to the point B, and the potential difference across R is denoted by V.

The arrowhead denotes that A is positive and the tail end of the arrow denotes that B is negative. Fig. 3 shows the convention of current and potential difference in a passive component, the resistor.

Fig. 3 Conventional current and voltage in a passive component.

Fig. 4 Conventional current and voltage in an active component.

The cell has an e.m.f. (electromotive force) E as shown in Fig. 4, the positive side is denoted by the arrowhead and the negative side is denoted by the tail end of the arrow. Observe that in Fig. 3, I and V are opposing each other and in Fig. 4, I and E are in the same direction. Note that current flows only if the circuit is closed. Referring to Fig. 2, we have $E - V = 0$ or $E - IR = 0$ or $E = V = IR$.

2

THE POTENTIAL DIFFERENCE (P.D.)

The potential difference between two points in a circuit is the electrical pressure or voltage required to cause the current flow between the points. The unit of potential difference is the volt, V. The *volt* is defined as the p.d. across a resistance of one ohm carrying a current of one ampere. The p.d. between two points is one volt if one joule of energy is used in sending one coulomb of charge between them.

$$1 \text{ volt} = \frac{1 \text{ joule}}{1 \text{ coulomb}}.$$

The unit of quantity of electric charge is the coulomb. One coulomb is the quantity of electricity conveyed by a current of one ampere flowing for one second.

OHM'S LAW

Ohm's Law states that the potential difference across a conductor is proportional to the current flowing through the conductor, provided the temperature is constant, that

is $V \propto I$ where $\propto$ is the proportionality sign and this sign is replaced by an equal sign by introducing a constant, R, in the equation

$V = RI$ where R = the resistance of the conductor.

$$R = \frac{V}{I} = \frac{\text{Potential difference across the conductor}}{\text{Current through the conductor}}.$$

LINEAR AND NON LINEAR RESISTORS

For linear resistors, the ratio of V over I $\left(\dfrac{V}{I}\right)$ is always the same and it is equal to the value of the linear resistance, R.

For non-linear resistors, the ratio of $\dfrac{V}{I}$ is different at different values of V or at different values of I. The following worked example is used to illustrate the difference between linear and non-linear resistors.

WORKED EXAMPLE 1

Tests on linear and non-linear resistors resulted in the following readings:-

V (volts)	0	15	25	35	45	60	
I (mA)	0	7.5	12.5	17.5	22.5	30 }	LINEAR
R (Ω)	( )	( )	( )	( )	( )	( ) }	
I (mA)	0	2	7.5	22.5	44	80 }	NONLINEAR
R (Ω)	( )	( )	( )	( )	( )	( ) }	

Complete the table and plot the following graphs:

(i) I against V for the linear and non linear resistors.

(ii) R against V for the linear and non linear resistors.

Using the above graphs draw your conclusions.

SOLUTION 1

(i)   LINEAR RESISTOR

$$R = \frac{V}{I} = \frac{15}{7.5 \times 10^{-3}} = 2000 \ \Omega \text{ at } V = 15 \text{ volts}$$

$$R = \frac{V}{I} = \frac{25}{12.5 \times 10^{-3}} = 2000 \ \Omega \text{ at } V = 25 \text{ volts}$$

$$R = \frac{V}{I} = \frac{35}{17.5 \times 10^{-3}} = 2000 \ \Omega \text{ at } V = 35 \text{ volts}$$

$$R = \frac{V}{I} = \frac{45}{22.5 \times 10^{-3}} = 2000 \ \Omega \text{ at } V = 45 \text{ volts}$$

$$R = \frac{V}{I} = \frac{60}{30 \times 10^{-3}} = 2000 \ \Omega \text{ at } V = 60 \text{ volts.}$$

Fig. 5 Graph of I against V, the gradient is the same everywhere and the relationship is linear and passes through the origin.

4

A graph of I (mA) is plotted against V (volts). It is linear and passes through the origin. The gradient is the same everywhere and the ratio $\dfrac{V}{I}$ is 2000 Ω. Even when the current through R is zero and V is zero, the resistance is 2000 Ω.

(ii)   NON LINEAR RESISTOR

At $V = 15$ volts, $R = \dfrac{V}{I} = \dfrac{15}{2 \times 10^{-3}}$ $= 7500 \ \Omega$

at $V = 25$ volts, $R = \dfrac{V}{I} = \dfrac{25}{7.5 \times 10^{-3}}$ $= 3333 \ \Omega$

at $V = 35$ volts, $R = \dfrac{V}{I} = \dfrac{35}{22.5 \times 10^{-3}}$ $= 1555 \ \Omega$

at $V = 45$ volts, $R = \dfrac{V}{I} = \dfrac{45}{44 \times 10^{-3}}$ $= 1024 \ \Omega$

at $V = 60$ volts, $R = \dfrac{V}{I} = \dfrac{60}{80 \times 10^{-3}}$ $= 750 \ \Omega.$

Fig. 6   Graph of I against V, the gradient at different points is different.
The relationship is non-linear.

The graph of I (mA) against V (volts) is non linear and the gradient at the five different values of V are not the same.

(i)   FOR THE LINEAR CASE

R (Ω) is plotted against V (volts) and is shown to be horizontal line, indicating that the resistance is constant.

5

Fig. 7 Graph of R against V. R is constant for all the values of V.

(ii) FOR THE NON LINEAR CASE

R (Ω) is plotted against V (volts) and is shown to be a curve, indicating that the resistance decreases as the voltage is increasing.

Fig. 8 Graph of R against V, R varies at different values of voltage.

SERIES RESISTORS

Three resistors in series can be connected as shown in Fig. 9, 10, 11 and 12.

Fig. 9 Resistors in series.

Fig. 10 Resistors in series

6

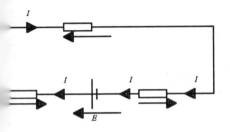

Fig. 11 Resistors in series.   Fig. 12 Resistors in series

In all the circuits in Fig. 9, 10, 11 and 12 above, it is observed that the current through each resistor is the same, therefore the resistors are in series. The sum of the p.d.s around a closed loop is zero or the sum of the voltages in a series circuit is equal to the total applied e.m.f.

$$V - IR_1 - IR_2 - IR_3 = 0$$

or $V = IR_1 + IR_2 + IR_3$

taking I as a common factor

$$V = I(R_1 + R_2 + R_3).$$

Let R_T be the total resistance of the circuit

$$V = IR_T$$

therefore $R_T = R_1 + R_2 + R_3$.

The total or equivalent resistance of a number of resistors connected in series is given by the sum of the individual resistances.

RESISTORS IN PARALLEL

Three resistors in parallel can be connected as shown in Fig. 13

Fig. 13

7

It is observed that the potential difference across each resistor is V, therefore the three resistors are in parallel.

The sum of the currents flowing through the three resistors connected in parallel is equal to the total current drawn from the supply.

$$I = I_1 + I_2 + I_3.$$

If R_T is the total resistance of the circuit, then $I = \dfrac{V}{R_T}$ and hence

$$\frac{V}{R_T} = \frac{V}{R_1} + \frac{V}{R_2} + \frac{V}{R_3} \text{ where } I_1 = \frac{V}{R_1}, I_2 = \frac{V}{R_2}, I_3 = \frac{V}{R_3}, \text{ according to ohm's}$$

law. Taking V as a common factor on the right hand side of the equation

$$\frac{V}{R_T} = V\left[\frac{1}{R_1} + \frac{1}{R_2} + \frac{1}{R_3}\right]$$

dividing both sides of the equation by V

$$\frac{1}{R_T} = \frac{1}{R_1} + \frac{1}{R_2} + \frac{1}{R_3} \quad \dots (1)$$

The equivalent resistance R_T, of three resistors connected in parallel is given by the equation (1).

If the number of resistors is extended to n then $\dfrac{1}{R_T} = \dfrac{1}{R_1} + \dfrac{1}{R_2} + \dfrac{1}{R_3} + \dots + \dfrac{1}{R_n}.$

For $n = 2$ $\quad \dfrac{1}{R_T} = \dfrac{1}{R_1} + \dfrac{1}{R_2}.$

It is convenient to re-arrange this formula as $\dfrac{1}{R_T} = \dfrac{R_2 + R_1}{R_1 R_2}$ or $R_T = \dfrac{R_1 R_2}{R_1 + R_2}.$

Thus the total resistance of two resistors in parallel is expressed as the product of the two resistors over their sum. This formula is *only* applicable for two resistors in parallel. This formula is useful to remember as

$$R_T = \frac{\text{product}}{\text{sum}}.$$

Re-arranging equation (1)

$$\frac{1}{R_T} = \frac{1}{R_1} + \frac{1}{R_2} + \frac{1}{R_3} = \frac{R_2 R_3 + R_1 R_3 + R_1 R_2}{R_1 R_2 R_3}$$

8

taking the reciprocal of each side, we have

$$R_T = \frac{R_1 R_2 R_3}{R_2 R_3 + R_1 R_3 + R_1 R_2}.$$

This formula is not so easy to remember, but useful for 3 resistors in parallel. Taking the reciprocal of each of equation (1)

$$R_T = \frac{1}{\dfrac{1}{R_1} + \dfrac{1}{R_2} + \dfrac{1}{R_3}}.$$

SERIES AND PARALLEL RESISTORS NETWORKS

WORKED EXAMPLE 2

Determine the total resistance of the networks between A and B showing clearly all the steps of your solution and expressing the results in three significant figures.

Network of resistors in series/parallel.

Fig. 14   Fig. 15

Fig. 16

Fig. 17

9

(v)

Fig. 18

(vi)

Fig. 19

(vii)

Fig. 20

SOLUTION 2

(i) $R_{AB} = R_1 + \dfrac{R_2 R_3}{R_2 + R_3} = 1 + \dfrac{2 \times 3}{2 + 3} = 1 + \dfrac{6}{5} = 2.2\ \Omega$

(ii) $R_{AB} = R_1 + \dfrac{R_2 R_3}{R_2 + R_3} + R_4 = 4 + \dfrac{5 \times 6}{5 + 6} + 7 = 4 + 2.73 + 7 = 13.7\ \Omega$

10

(iii) $R_{AB} = \dfrac{R_1 R_2}{R_1 + R_2} + R_3 + \dfrac{R_4 R_5 R_6}{R_4 R_5 + R_4 R_6 + R_5 R_6}$

$\quad = \dfrac{8 \times 9}{8 + 9} + 10 + \dfrac{12 \times 6 \times 12}{72 + 72 + 144} = 4.24 + 10 + 3 = 17.2 \ \Omega.$

An easier way to find the combination of the three resistors in parallel is to use the fact that, the total resistance of two equal resistors in parallel is equal to the half value of one, thus 12 Ω in parallel with 12 Ω is a 6 Ω resistor then 6 Ω in parallel with the third 6 Ω resistor gives the combination 3 Ω.

(iv) $R_{AB} = \dfrac{R_1 R_2 R_3}{R_1 R_2 + R_1 R_3 + R_2 R_3} + \dfrac{R_4 R_5}{R_4 + R_5} + \dfrac{R_6 R_7 R_8}{R_6 R_7 + R_7 R_8 + R_6 R_8}$

$R_{AB} = \dfrac{4 \times 2 \times 1}{4 \times 2 + 4 \times 1 + 2 \times 1} + \dfrac{4 \times 2}{4 + 2} + \dfrac{7 \times 10 \times 15}{7 \times 10 + 7 \times 15 + 10 \times 15}$

$R_{AB} = 0.571 + 1.33 + 3.23 = 5.13 \ \Omega$

$R_{AB} = 5.13 \ \Omega.$

(v) $R_{AB} = R_1 + \dfrac{R_2 R_3}{R_2 + R_3} + \dfrac{R_4 R_5 R_6}{R_4 R_5 + R_4 R_6 + R_5 R_6} +$

$$\dfrac{R_7 R_8 R_9 R_{10}}{R_7 R_8 R_9 + R_7 R_8 R_{10} + R_7 R_9 R_{10} + R_8 R_9 R_{10}}$$

$R_{AB} = 1 + \dfrac{2 \times 3}{2 + 3} + \dfrac{4 \times 5 \times 6}{4 \times 5 + 4 \times 6 + 5 \times 6} +$

$$\dfrac{7 \times 8 \times 9 \times 10}{7 \times 8 \times 9 + 7 \times 8 \times 10 + 7 \times 9 \times 10 + 8 \times 9 \times 10}$$

$R_{AB} = 1 + 1.2 + 1.62 + 2.09 = 5.91 \ \Omega.$

The four parallel resistors can be found easier as follows:-

$\dfrac{1}{R_T} = \dfrac{1}{R_7} + \dfrac{1}{R_8} + \dfrac{1}{R_9} + \dfrac{1}{R_{10}} = \dfrac{1}{7} + \dfrac{1}{8} + \dfrac{1}{9} + \dfrac{1}{10}$

$\quad = 0.143 + 0.125 + 0.111 + 0.1 = 0.479$

taking the reciprocals, we have
$$R_T = 2.09 \ \Omega.$$

11

(vi) $\quad R_{T_1} = \dfrac{R_2 R_3}{R_2 + R_3} \qquad R_{T_2} = \dfrac{R_4 R_5 R_6}{R_4 R_5 + R_4 R_6 + R_5 R_6} \qquad R_T = R_{T_1} + R_{T_2}$

$$R_1 + \dfrac{\left(R_{T_1} + R_{T_2}\right)R_7}{R_{T_1} + R_{T_2} + R_7} = R = \dfrac{R_T R_7}{R_T + R_7} + R_1 .$$

Fig. 21 Network of resistors in series/parallel.

$$R_{T_1} = \dfrac{R_2 R_3}{R_2 + R_3} = \dfrac{50 \times 25}{50 + 25} = 16.7 \ \Omega$$

$$R_{T_2} = \dfrac{R_4 R_5 R_6}{R_4 R_5 + R_4 R_6 + R_5 R_6} = \dfrac{68 \times 37 \times 68}{68 \times 37 + 68 \times 68 + 37 \times 68} = 17.7 \ \Omega$$

$$R_T = R_{T_1} + R_{T_2} = 16.7 + 17.7 = 34.4 \ \Omega$$

$$R = \dfrac{R_T R_7}{R_T + R_7} + R_1 = \dfrac{34.4 \times 10}{34.4 + 10} + 30 = 7.75 + 30 = 37.8 \ \Omega$$

$$R = 37.8 \ \Omega$$

$$R_{AB} = \dfrac{R R_8}{R + R_8} = \dfrac{37.8 \times 100}{37.8 + 100} = 27.4 \ \Omega.$$

(vii) $\quad R_{T_1} = \dfrac{R_2 R_3}{R_2 + R_3} \qquad R_{T_2} = \dfrac{R_4 R_5}{R_4 + R_5} \qquad R_T = R_{T_1} + R_{T_2}$

$$R = \dfrac{R_T R_6}{R_T + R_6} \qquad R_{AB} = \dfrac{\left(R_1 + R\right)R_7}{R_1 + R + R_7}$$

$$R_{T_1} = \dfrac{R_2 R_3}{R_2 + R_3} = \dfrac{2 \times 4}{2 + 4} = \dfrac{4}{3} = 1.33 \ \Omega$$

$$R_{T_2} = \frac{R_4 R_5}{R_4 + R_5} = \frac{12 \times 6}{12 + 6} = \frac{72}{18} = 4\ \Omega$$

$$R_T = R_{T_1} + R_{T_2} = 1.33 + 4 = 5.33\ \Omega$$

$$R = \frac{R_T R_6}{R_T + R_6} = \frac{5.33 \times 3}{5.33 + 3} = 1.92\ \Omega$$

$$R_{AB} = \frac{(2.08 + 1.92)\,16}{2.08 + 1.92 + 16} = 3.2\ \Omega$$

b.   *Derives and applies the concept of proportional voltage and current division in circuit analysis.*

THE CURRENT DIVIDER

Fig. 22   Current divider. Equivalent circuit.

The total resistance for the circuit is

$$R_T = \frac{R_1 R_2}{R_1 + R_2}$$

$$\frac{V}{I} = \frac{R_1 R_2}{R_1 + R_2} \qquad\qquad V = \frac{R_1 R_2}{R_1 + R_2}I = \frac{(R_1 R_2)I}{R_1 + R_2}$$

$$I_1 = \frac{V}{R_1} = \frac{(R_1 R_2)I}{R_1 + R_2}\left[\frac{1}{R_1}\right] = \frac{R_2}{R_1 + R_2}I$$

$$\boxed{I_1 = \frac{R_2}{R_1 + R_2}I} \quad \dots (1)$$

13

$$I_2 = \frac{V}{R_2} = \frac{(R_1 R_2)I}{R_1 + R_2}\left[\frac{1}{R_2}\right] = \frac{R_1}{R_1 + R_2}I$$

$$\boxed{I_2 = \frac{R_1}{R_1 + R_2}I} \quad \ldots (2)$$

WORKED EXAMPLE 3

Calculate the currents in the 1 Ω and 2 Ω resistors.

Fig. 23 Current divider.

SOLUTION 3

The total resistance of the circuit

$$R_T = \frac{R_1 R_2}{R_1 + R_2} = \frac{1 \times 2}{1 + 2} = \frac{2}{3}\ \Omega.$$

The total current $I = \dfrac{V}{R_T} = \dfrac{10}{2/3} = \dfrac{30}{2} = 15$ A.

Using equations (1) and (2)

we have $I_1 = \dfrac{2}{3}(15) = 10$ A

$$I_2 = \frac{1}{3}(15) = 5 \text{ A}.$$

WORKED EXAMPLE 4

The total current in Fig. 22, $I = 2$ A. If $R_1 = 10$ KΩ and $R_2 = 30$ KΩ, determine the currents I_1 and I_2.

14

SOLUTION 4

$$I_1 = \frac{R_2}{R_1 + R_2}I = \frac{30 \times 10^3}{(10 + 30) \times 10^3} \times 2 = \frac{60}{40} = 1.5 \text{ A}$$

$$I_2 = 2 - 1.5 = 0.5 \text{ A}$$

Fig. 24 Current divider.

WORKED EXAMPLE 5

In Fig. 22, if $I_1 = 5$ A, $R_2 = 10\ \Omega$ and $I = 12$ A, determine the value of R_1.

SOLUTION 5

$$I_2 = I - I_1 = 12 - 5 = 7 \text{ A}$$

p.d. across $R_2 = I_2 R_2 = 7 \times 10 = 70$ volts

p.d. across $R_1 = 70 = I_1 R_1$

$$R_1 = \frac{70}{5} = 14\ \Omega.$$

Fig. 25 Current divider

15

THE POTENTIAL DIVIDER

The potential divider or potentiometer is used in order to obtain a variable voltage from a constant voltage supply.

Consider the circuit in Fig. 26. The input voltage is fixed (V) and the output voltage (V_1) is taken at the terminals A and B.

Fig. 26 Potential divider.

By changing the position of the sliding contact A, we can obtain any voltage from zero to the value of V.

$V = I(R_1 + R_2)$ where the sliding contact divides the resistance R into resistances R_1 and R_2

$V_1 = IR_2$

$V = I(R_1 + R_2)$ ... (3)

$V_1 = IR_2$ ... (4)

Dividing equation (3) by equation (4)

$$\frac{V}{V_1} = \frac{I(R_1 + R_2)}{IR_2}$$

we have $\dfrac{V}{V_1} = \dfrac{R_1 + R_2}{R_2}$

or $$\boxed{V_1 = \frac{R_2}{R_1 + R_2} V}$$

16

WORKED EXAMPLE 6

If $V = 100$ volts, $R_2 = 10$ KΩ, $R_1 = 90$ KΩ, calculate V_1, the p.d. across R_2, and V_2 across R_1.

SOLUTION 6

$$V_1 = \frac{10 \times 10^3}{10 \times 10^3 + 90 \times 10^3} \times 100 = \frac{10}{100} \times 100 = 10 \text{ volts}$$

$$V = V_1 + V_2 \Rightarrow V_2 = V - V_1 = 100 - 10 = 90 \text{ volts}$$

Fig. 27 Potential divider

WORKED EXAMPLE 7

The circuit diagram of a potential divider of 100 Ω is as shown in Fig. 28.

AB is connected across a 100 V d.c. supply and CD is connected across a 5 Ω resistor which takes a current of 1 A. Determine the position of the sliding contact, X.

Fig. 28 Potential divider

SOLUTION 7

Let R_1 and R_2 be the resistances of the 100 Ω. The total resistance of the circuit is

$$R_T = R_1 + \frac{R_2 R_L}{R_2 + R_L}.$$

The total current, $I = \dfrac{V}{R_T}$.

Fig. 29 Loaded potential divider.

Let the resistance of $AX = R_1$ then $BX = R_2 = 100 - R_1$

$$\frac{100}{R_1 + \dfrac{(100 - R_1)5}{100 - R_1 + 5}} = I$$

the current through $R_L = 1 = \dfrac{R_2}{R_2 + R_L}I = \dfrac{100 - R_1}{100 - R_1 + R_L}I$

therefore $\dfrac{100 - R_1}{100 - R_1 + 5}I = 1$

therefore $\dfrac{100 - R_1}{100 - R_1 + 5} \cdot \dfrac{100}{R_1 + \dfrac{(100 - R_1)5}{100 - R_1 + 5}} = 1$

$$\frac{(100 - R_1)100(100 - R_1 + 5)}{(100 - R_1 + 5)\left[R_1(100 - R_1 + 5) + 5(100 - R_1)\right]} = 1$$

$$100(100 - R_1) = R_1(100 - R_1 + 5) + 5(100 - R_1)$$

$$10000 - 100R_1 = 100R_1 - R_1^2 + 5R_1 + 500 - 5R_1$$

$$R_1^2 - 200R_1 + 9500 = 0$$

18

$$R_1 = \frac{200 \pm \sqrt{200^2 - 4 \times 9500}}{2} = \frac{200 \pm 44.72}{2}$$

$R_1 = 122.36 \ \Omega$ or $R_1 = 77.64 \ \Omega$.

The former is disregarded, and therefore $R_1 = 77.64 \ \Omega \approx 78 \ \Omega$ and $R_2 \approx 22 \ \Omega$.

WORKED EXAMPLE 8

Calculate R_1 in the network shown in Fig. 30.

Fig. 30 Loaded potential divider

SOLUTION 8

The p.d. across the 20 Ω resistor = $20 \times 2 = 40$ V therefore the p.d. across $R_1 = 240 - 40 = 200$ V the p.d. across the $160 - R_1$ resistor is 40 V the current through the $(160 - R_1)$ resistor is

$$I_1 = \frac{40}{160 - R_1}$$

$$I = I_1 + 2 = \frac{40}{160 - R_1} + 2 = \frac{200}{R_1}$$

$$\frac{40R_1}{160 - R_1} + 2R_1 = 200$$

$$40R_1 + 2R_1(160 - R_1) = 200(160 - R_1)$$

$$40R_1 + 320R_1 - 2R_1^2 = 32000 - 200R_1$$

$$2R_1^2 - 200R_1 - 360R_1 + 32000 = 0$$

$$R_1^2 - 280R_1 + 16000 = 0$$

$$R_1 = \frac{280 \pm \sqrt{280^2 - 4 \times 16000}}{2} = \frac{280 \pm 120}{2}$$

$$R_1 = \frac{400}{2} = 200 \ \Omega \qquad \text{or} \qquad R_1 = \frac{160}{2} = 80 \ \Omega$$

$R_1 = 80 \ \Omega$ is the correct value as 160 − 200 produces a negative resistance.

SUMMARY 1

OHM'S LAW

THE POTENTIAL DIFFERENCE ACROSS A CONDUCTOR IS PROPORTIONAL TO THE CURRENT FLOWING THROUGH THE CONDUCTOR, PROVIDED THE TEMPERATURE IS CONSTANT.

CURRENT DIVIDER

THE CURRENT THROUGH ONE RESISTOR OF TWO RESISTORS IN PARALLEL IS EQUAL TO THE RATIO OF THE OTHER RESISTOR OVER THE SUM OF THE TWO RESISTORS TIMES THE TOTAL CURRENT.

$$I_1 = \frac{R_2}{R_1 + R_2} I$$

$$I_2 = \frac{R_1}{R_1 + R_2} I$$

where I_1 is the current through R_1, I_2 is the current through R_2 and I is the total current.

VOLTAGE DIVIDER

THE VOLTAGE ACROSS A RESISTOR OF RESISTANCE R_1 OF TWO RESISTORS IN SERIES IS EQUAL TO THE RATIO OF R_1 TO $R_1 + R_2$ TIMES THE TOTAL VOLTAGE

$$V_1 = \frac{R_1}{R_1 + R_2} V$$

$$V_2 = \frac{R_2}{R_1 + R_2} V$$

where V_1 is the p.d. across R_1, V_2 is the p.d. across R_2, and V is the total voltage.

EXERCISES 1

1. Three resistors of 1, 2 and 3 Ω are connected in parallel. If the total current taken is 12 A find the current through each resistor.
    (Ans. 6.55A, 3.27 A, 2.18 A).

2. Two resistors of 3 Ω and 5 Ω are connected in parallel. If the total current taken is 1 A find the current through each resistor.
    (Ans. 0.625 A, 0.375 A)

3. Two resistors of 2 Ω and 8 Ω are connected in parallel. If the total current taken is 5 A find the current through each resistor.
    (Ans. 1 A, 4 A)

4. Determine the currents, I_1, and I_2 shown in Fig. 31.

Fig. 31 Current divider

(Ans. 1.2 A, 2.4 A)

5. Determine the currents I_1, I_2 and I_3 as shown in Fig. 32.

Fig. 32 Currents in parallel circuit

(Ans. 0.85 A, 0.64 A, 0.51 A)

6. If the total resistance of the network is 1 Ω, in Fig. 33.
    (i)   Find the value of the resistance, R.
    (ii) Find the current I_1 and I_2 and the total current taken from the supply, I.

Fig. 33 Current in parallel circuit

(Ans. (i) 1.91 Ω (ii) 6.37 A, 2.73 A, 19.1 A).

7. The total resistance of this combination of resistors is 10 Ω, in Fig. 34. Determine the value of R_1, R_2, R_3 and R_4.

Fig. 34 Resistor in parallel circuit

(Ans. 100 Ω, 50 Ω, 33.3 Ω, 25 Ω)

8. The resistance of a moving coil instrument is 5 Ω and it is provided with a shunt of resistance 0.0005 Ω. Find the current in the instrument when used to measure 15 A.
   (Ans. 1.5 mA)

9. A resistor R is connected across a 100 V supply. A voltmeter of resistance 1500 Ω is connected between the centre of the resistor and one side of the supply. Determine R if the voltmeter reading is 35 V.
   (Ans. 2570 Ω)

10. Two resistors are connected in series as shown in Fig. 35 and Fig. 36 and the voltmeters V_1 and V_2 are connected in turn to measure the p.d. across R_1 and across R_2.

   Determine the values of R_1 and R_2, if V_1 = 80 volts and V_2 = 100 volts.

Fig. 35 The effect of the resistance of a voltmeter

Fig. 36 The effect of the resistance of a voltmeter

(Ans. R_1 = 2000 Ω, R_2 = 2500 Ω)

11. Determine the value of the potential difference across the 5 Ω resistor, V_1 in Fig. 37. (Ans. 1.67 V)

Fig. 37 Potential divider

12. Determine the value of the p.d. across the two 10 KΩ resistor, V_1 as shown in Fig. 38. (Ans. 16.7 V)

Fig. 38 Loaded potential divider

23

13. Determine the value of V by employing

    (i)   the potential divider method

    (ii)   the current divider method in Fig. 39. (Ans. 4.29 volts)

Fig. 39 Ladder network. Potential divider principle.

14. Fig. 40 shows a d.c. source connected across a variable load resistance R. The e.m.f. of the source is 6 V, its internal resistance is 1 Ω. The table below shows the current through the load in amperes, the load resistance in ohms and the power dissipated in the load in watts.

I (A)	6	( )	4	( )	3	( )	2.4	( )	2
R (Ω)	0	$\frac{1}{4}$	$\frac{1}{2}$	$\frac{3}{4}$	1	$1\frac{1}{4}$	$1\frac{1}{2}$	$1\frac{3}{4}$	2
P (W)	0	( )	8	( )	9	( )	8.64	( )	8

Complete the table and plot P against R and from your graph determine the maximum power dissipated in the load and note the value of R where this power occurs. Draw your conclusions. (Ans. 9 W, $R = r = 1$ Ω)

Fig. 40 Maximum power transfer

15. Fig. 41 shows a series-parallel resistor network with the given data.

Fig. 41 Loaded potential divider

Calculate the value of R showing clearly the steps of your solution.
(Ans. 6 Ω).

16. (a) Determine the total resistance of the network shown in Fig. 42.

Fig. 42 Network of resistors. Determination of currents

(b) If the p.d. across AB is 10 volts.

Determine: (i) the total current I_1,

(ii) the current, I_2,

(iii) the current, I_3.

(Ans. (a) 2.5 Ω (b) (i) 4 A (ii) 2 A (iii) 1.33 A.)

2. KIRCHHOFF'S LAWS

c. Applies Kirchhoff's Laws to problems involving not more than two unknowns.

NODE OR JUNCTION OF A NETWORK

A node or a junction is shown in Fig. 43 as the bold dot at N. Currents flow into, or out of the node or junction.

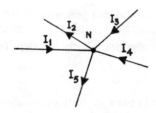

Fig. 43 Many wires connected to a common point (node or junction)

CONNECTION OF CURRENT FLOW

If the currents flowing into the node are assumed to be positive then the currents flowing out of the node are negative. If the currents flowing into the node are assumed to be negative then the currents flowing out of the node are positive.

KIRCHHOFF'S CURRENT LAW (K.C.L.)

The sum of the currents flowing at the junction is equal to zero. Referring to Fig. 43

$$I_1 + I_3 + I_4 - I_2 - I_5 = 0$$

or $- I_1 - I_3 - I_4 + I_2 + I_5 = 0.$

We can clearly see that it does not matter which currents are positive and which currents are negative. If I_1, I_3, I_4 are positive then I_2, I_5 are negative and if I_1, I_3, I_4 are negative then I_2, I_5 are positive.

LOOP OR MESH OF A NETWORK

A simple loop or mesh is shown in Fig. 44

Fig. 44 One loop or mesh

where there is continuity of current and the loop is closed. It starts from the positive pole of the e.m.f., E, through the resistor R, the load, the internal resistance, r, of the cell, and finishes at the negative pole of the e.m.f.

VOLTAGE AND CURRENT CONVENTIONS

The current I is assumed to flow from the positive pole of E to the negative pole of E. The p.d.s established in the network of Fig. 44 are one across E, one across R and one across r. Fig. 45 shows the first p.d. across the cell which is an active component.

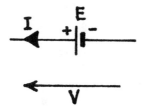

Fig. 45 Conventional current and voltage in active component.

The positive terminal is shown by the arrow head of the voltage arrow and the negative terminal is shown by the tail end of the arrow.

I and V in the active device are in the same direction. The other voltages are shown across the passive components, R and r.

Fig. 46 shows the current flowing through the passive component from A to B, and since the current flows from A to B, A is more positive than B.

Fig. 46 Conventional current and voltage in a passive component.

I and *V* are in opposition.

Similarly the p.d. across *r* is as shown in Fig. 47. *I* flows from C to D, and C is more positive than D the arrow head shows the more positive potential.

Fig. 47 Conventional current and voltage in a passive component.

The p.d.s around the closed loop of Fig. 48 may be written as $E - IR - Ir = 0$ ... (2) considering the anticlockwise p.d., E to be positive and clockwise p.d.s Ir and IR to be negative.

Fig. 48 Conventional currents and voltages in a one loop or mesh circuit.

Alternatively, if we consider the clockwise p.d. positive then the anticlockwise p.d. is negative. This is shown if we change all the sign of equation (2)

$$- E + IR + Ir = 0$$

therefore $$Ir + IR = E$$

or $$I = \frac{E}{r + R}$$

28

KIRCHHOFF'S VOLTAGE LAW (K.V.L.)

The sum of the e.m.f.s around the closed loop is ZERO. Referring to Fig. 48.

$$E - IR - Ir = 0$$

or $\quad E = I(R + r)$.

It is very important for the student to practice and understand the conventions of currents and voltages for active and passive components.

For this reason, the student mark the currents and the p.d.s for various networks shown below.

WORKED EXAMPLE 9

Determine the current in each branch and the magnitude and direction of the p.d. between points A and C, for the circuit in Fig. 49.

Determine also, the values V_{AD} and V_{ED} and hence V_{AE}.

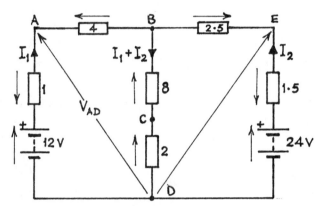

Fig. 49 Conventional currents and voltages in a two loop or mesh circuit.

SOLUTION 9

Assume the currents flow as shown.

Applying *KVL* to loops *ABDA* and *BEDB*.

For *ABDA* loop

$$12 - I_1(1) - 4I_1 - 8(I_1 + I_2) - 2(I_1 + I_2) = 0$$

or $\quad 15I_1 + 10I_2 = 12$

or $\quad \boxed{3I_1 + 2I_2 = 2.4} \quad \ldots (1)$

29

For *BEDB* loop

$$2(I_1 + I_2) + 8(I_1 + I_2) + 2.5I_2 + 1.5I_2 = 24$$

or $\quad 10I_1 + 14I_2 = 24$

or $\quad \boxed{I_1 + 1.4I_2 = 2.4} \quad$ ... (2)

Solving the simultaneous equation (1) and (2)

$$3I_1 + 2I_2 \quad = 2.4$$
$$I_1 + 1.4I_2 = 2.4 \quad \times (-3)$$

$$3I_1 + 2I_2 \quad = 2.4 \qquad \text{Adding}$$
$$-3I_1 - 4.2I_2 = -7.2$$
$$-2.2I_2 = -4.8$$

$$I_2 = \frac{2.4}{1.1} = 2.18 \text{ A.}$$

From equation (2), $I_1 + 1.4(2.18) = 2.4$

$$I_1 = -0.652 \text{ A}$$

$V_{AC} = 4I_1 + 8(I_1 + I_2) = 4(-0.652) + 8(-0.652 + 2.18)$
$V_{AC} = -2.608 + 12.224$
$\underline{V_{AC} = 9.62 \text{ volts.}}$

$V_{AD} = 4I_1 + 10(I_1 + I_2) = 14I_1 + 10I_2$
$V_{AD} = 14(-0.652) + 10 \times 2.18 = -9.128 + 21.8$
$\underline{V_{AD} = 12.7 \text{ volts.}}$

$V_{ED} = 2.5I_2 + 10(I_1 + I_2) = 10I_1 + 12.5I_2$
$V_{ED} = 10I_1 + 12.5I_2 = 10(-0.652) + 12.5(2.18) = -6.52 + 27.25$
$\underline{V_{ED} = 20.7 \text{ volts.}}$

$V_{AE} + V_{ED} = V_{AD}$
$V_{AE} = V_{AD} - V_{ED} = 12.7 - 20.7$
$V_{AE} = -8 \text{ volts.}$

therefore E is positive with respect to A.

WORKED EXAMPLE 10

The direction of branch currents are assumed in the following circuit diagrams and the direction of potential differences across each component are also shown.

SOLUTION 10

Fig. 50

Fig. 51

Fig. 52   Fig. 53

Fig. 54   Fig. 55

Fig. 56   Fig. 57

Fig. 58   Fig. 59

Fig. 60

Fig. 61

WORKED EXAMPLE 11

Use Kirchhoff's Laws to calculate (a) the three branch currents in the circuit of Fig. 62
(b) the potential difference between A and E, stating which point is at the higher potential.

Fig. 62

SOLUTION 11

(a) Let the currents be as shown in the diagram of Fig. 63. The p.d. across each component is marked by a large arrow.
Applying Kirchhoff's laws for the loops: ABCD and BEFCB, we have

Fig. 63

ABCDA loop

$$80 = 40I_1 + 120(I_1 - I_2)$$

or $\quad 160I_1 - 120I_2 = 80$

or $\quad \boxed{4I_1 - 3I_2 = 2} \qquad ... (1)$

BEFCB loop

$$120(I_1 - I_2) - 60I_2 = -30$$

or $\quad 120I_1 - 180I_2 = -30$

or $\quad \boxed{4I_1 - 6I_2 = -1} \qquad ... (2)$

Solving the simultaneous equations (1) and (2)

$$4I_1 - 3I_2 = 2 \qquad ... (1)$$
$$4I_1 - 6I_2 = -1 \qquad ... (2)$$

Equation (1) minus equation (2)

$$3I_2 = 3$$
$$I_2 = 1A.$$

Substituting in equation (1) this value

$$4I_1 - 3(1) = 2$$
$$4I_1 = 5$$
$$I_1 = 1.25 \text{ A.}$$

The three branch currents through 40 Ω, 60 Ω and 120 Ω are: 1.25 A, 1.00 A and 0.25 A.

(b) $V_{AE} = 40I_1 + 60I_2 = 40(1.25) + 60(1) = 50 + 60$

$V_{AE} = 110$ volts

A is at a higher potential than E.

WORKED EXAMPLE 12

Use Kirchhoff's laws to obtain the values of the currents I_1, I_2 and I_3 in Fig. 64.

Fig. 64

SOLUTION 12

Mark the voltages across the passive and active components by large arrows, in the Fig. 65.

Fig. 65

Loop *ABCDA*

Taking a clockwise direction around this loop

$$36 + 60 - 4I_1 - 12I_2 = 0$$

or $4I_1 + 12I_2 = 96$

or $\boxed{I_1 + 3I_2 = 24}$ ... (1)

Loop *BFECB*

Taking again a clockwise direction around this loop

$$12I_2 - 3I_3 - 60 = 0 \text{ but } I_3 = I_1 - I_2$$

or $\qquad 12I_2 - 3I_1 + 3I_2 = 60$

or $\qquad -3I_1 + 15I_2 = 60$

or $\qquad \boxed{-I_1 + 5I_2 = 20} \quad \dots (2)$

Solving the simultaneous equations

$$I_1 + 3I_2 = 24 \quad \dots (1)$$

$$-I_1 + 5I_2 = 20 \quad \dots (2)$$

adding (1) and (2) we have

$$8I_2 = 44$$

$$I_2 = \frac{22}{4} = \underline{5.5 \text{ A.}}$$

From (1) $\qquad I_1 + 3(5.5) = 24$

$$I_1 = 24 - 16.5 = 7.5 \text{ A}$$

$$\underline{I_1 = 7.5 \text{ A}}$$

$$I_3 = I_1 - I_2 = 7.5 - 5.5 = 2.0 \text{ A}$$

$$\underline{I_3 = 2.0 \text{ A.}}$$

WORKED EXAMPLE 13

(a)   (i)   State the number of junctions or nodes that exist in the circuit of Fig. 66.

       (ii)   Assuming that the branch currents are flowing as shown in the circuit of Fig. 66, apply Kirchhoff's current law and write down an equation at one of the junctions.

       (iii)   State and label the number of loops that exist in the circuit of Fig. 66. Apply Kirchhoff's voltage law, and write down two equations of two such loops, in terms of I_1 and I_2.

       (iv)   Solve the two simultaneous equations you have obtained in (ii) and (iii) giving your answers in three significant figures.

       (v)   Redraw the circuit of Fig. 66 and indicate the branch currents and p.d.s by small and large arrows.

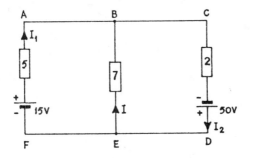

Fig. 66

(b) For the circuit shown in Fig. 67 calculate the resistance between the terminals
     A and B.

Fig. 67

SOLUTION 13

(a) (i) There are two junctions or nodes, namely the points B and E

(ii) At B
$$I_1 + I - I_2 = 0$$

At E
$$I = I_2 - I_1$$

(iii) Three loops

$ABEFA$, $BCDEB$, and $ABCDEFA$.

Two such loops are:

$ABEFA$ loop
$$15 + 7I = 5I_1$$
$$15 + 7(I_2 - I_1) - 5I_1 = 0$$

$$\boxed{-12I_1 + 7I_2 = -15}$$

37

BCDEB loop

$50 = 2I_2 + 7I$

$7(I_2 - I_1) + 2I_2 = 50$

$$\boxed{-7I_1 + 9I_2 = 50}$$

(iv) Solving the equation (1) and (2) simultaneously

$- 7I_1 + 9I_2 = 50 \qquad \ldots (1) \times (12)$

$- 12I_1 + 7I_2 = - 15 \qquad \ldots (2) \times (- 7)$

Multiplying (1) by 12 and (2) by (– 7) we have

$- 84I_1 + 108I_2 = 600 \qquad \ldots (3)$

$84I_1 - 49I_2 = 105 \qquad \ldots (4)$

Adding equations (3) and (4)

$59I_2 = 705$

$I_2 = 11.95$ A

substituting in (1)

$- 7I_1 + 9(11.95) = 50$

$- 7I_1 = 50 - 107.55$

$I_1 = 8.22$ A

$I = I_2 - I_1 = 11.95 - 8.22 = 3.73$ A

$I = 3.73$ A.

(v) The direction of currents are correct as shown in Fig. 68

Fig. 68

(b)   Mark the terminals as shown in Fig. 69.

Fig. 69

R_1 has a terminal A and a terminal B
R_2 has a terminal A and a terminal B
R_3 has a terminal A and a terminal B

therefore the three resistors are re-drawn as shown in Fig. 70, showing three common terminals A and three common terminals B.

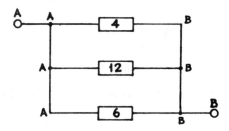

Fig. 70

$$R_{AB} = \frac{R_1 R_2 R_3}{R_1 R_2 + R_1 R_3 + R_2 R_3} = \frac{4 \times 12 \times 6}{4 \times 12 + 4 \times 6 + 12 \times 6} = 2\ \Omega$$

or $\dfrac{1}{R_{AB}} = \dfrac{1}{4} + \dfrac{1}{12} + \dfrac{1}{6} = 0.25 + 0.0833 + 0.1666 = 0.5$

or $R_{AB} = \dfrac{1}{0.5} = 2\ \Omega.$

SUMMARY 2

KIRCHHOFF'S CURRENT LAW (K.C.L.)

THE SUM OF THE CURRENTS FLOWING AT A JUNCTION OR NODE IS EQUAL TO ZERO

KIRCHHOFF'S VOLTAGE LAW (K.V.L.)

THE SUM OF THE EMF'S AROUND A CLOSED LOOP IS EQUAL TO ZERO.

EXERCISES 2

1. Using Kirchhoff's Laws calculate the current in each branch of the circuit of Fig. 71.

Fig. 71 Kirchhoff's Law

(Ans. 1.175 A, − 0.41 A, 0.765 A)

2. Using Kirchhoff's Laws, calculate:

   (a) the three branch currents in the circuit of Fig. 72

Fig. 72 Kirchhoff's Law

   (b) the p.d. between the points P and Q, and

       the p.d. between the points X and Y.

(Ans. (a) 81.7 A, 1.30 A, 83.0 A (b) 10 V, 10 V)

3. Calculate the current in each circuit when the tapping is set such that $AC = 25 \ \Omega$, in Fig. 73

Fig. 73 Kirchhoff's Laws.

(Ans. 5.63 A, - 4.82 A)

4. Apply Kirchhoff's current Law (K.C.L.) to the junction A and Kirchhoff's Voltage Law (K.V.L.) to the loops (1) and (2) of the circuit, of Fig. 74, and hence find I_1 and I_2.

Fig. 74 Kirchhoff's Laws

(Ans. - 1.87 A, 4.06 A)

5. Determine the currents I_1, I_2 and I_L in the circuit of Fig. 75

Fig. 75 Kirchhoff's Laws

(Ans. I_1 = 7.86 A, I_2 = - 6.91 A, I_L = 0.95 A)

41

6. Calculate the current in the network of Fig. 50.

(Ans. 5.83 A)

7. Referring to the circuit of worked example 10 Fig. 52, determine the currents I_1, I_2 and I_3 which are assumed to flow from the positive terminals or poles of the cells 2 V, 3 V and 5 V respectively. Redraw the circuit by inserting the actual flow of currents. State the discharging and charging currents.

(Ans. − 0.3125 A, 0.075 A, 0.2375 A)

8. Calculate the current through the load of 100 Ω in the circuit of Fig. 53.

(Ans. − 0.1 A)

9. Determine the magnitude and direction of the load current in the circuit of Fig. 54 and the p.d. across the load.

(Ans. 0.04 A from B to A, 0.60 V)

10. Determine the actual currents through the 50 V, 20 V and 100 V d.c. supplies of Fig. 57 redraw the circuit indicating the actual currents flowing.

(Ans. − 44 A, − 14.8 A, 58.8 A)

11. Determine the currents through the resistors 1 Ω, 5 Ω and 15 Ω and the p.d. across A and B of Fig. 58.

(Ans. 5.26 A, 3.95 A, 1.32 A, 19.73 V)

12. Calculate the currents through the 100 V and 200 V d.c. supplies of Fig. 59.

(Ans. 48.5 A, − 25.8 A)

13. Determine the p.d. across A and B of Fig. 61

(Ans. 22.14 V)

14. Determine the equivalent resistance between A and G, the diagonal of the cube or resistors of Fig. 60, employing Kirchhoff's Laws.

$$\left(\text{Ans. } \frac{5}{6} \ \Omega \right).$$

15. Two batteries of e.m.f.s of 5 V and 10 V have internal resistances of 1 Ω and 5.5 Ω respectively are connected in parallel across a load of 4.5 Ω resistance. Fig. 76. Determine the branch currents.

Fig. 76

(Ans. 0.144 A, 0.935 A, 1.079 A)

16. Two batteries of e.m.f.s of 110 V and 120 V have internal resistances of 20 Ω and 40 Ω respectively are connected in parallel across a load of 100 Ω. Determine the p.d. across the load.

Fig. 77

(Ans. 100 V)

17. Two batteries of e.m.f.s of 10 V and 5 V have internal resistances of 1 Ω and 1.5 Ω respectively are connected in parallel across a 10 Ω local. Determine the load current.

Fig. 78

(Ans. I = 0.75 A)

43

3. ELECTRIC FIELDS

3. Applies the fundamental laws and properties of electric fields to problems involving capacitors

   a. Introduces the concepts of electric field and electric flux to explain the forces of attraction and repulsion between charged bodies and defines electric field strength, potential and potential difference in terms of force and work done on a unit charge.

COULOMB'S LAW

The mechanical force, F, in newtons exerted between two charged bodies is given by the formula

$$F = \frac{Q_1 Q_2}{4\pi \in r^2}$$

where Q_1 and Q_2 are the charges of the bodies, r is the distance between the charges and $\in$ is the permittivity of the medium.

If the charges are unlike, the force is *attractive* and if the charges are like, the force is *repulsive*. This formula describes Coulomb's law, the force is proportional to the product of the charges, and the force is inversely proportional to the square of their distance apart, r. If the medium is vacuum or free space then $\in$ is $\in_o$ and is the permittivity of free space ($\in_o$ = 8.85 × 10^{-12} F/m). The force is reduced appreciably if the permittivity of the medium is high.

ELECTRIC FIELDS

A charged sphere has an electric field, that is represented by a field of radial electric lines as shown in Fig. 79. Two parallel plates with positive charge on one plate has an electric field, represented by parallel lines as shown in Fig. 80.

Fig. 79   Fig. 80

The region around the electric charge Q constitutes an electric field because any other charge placed in that region experiences a mechanical force given by Coulomb's law.

Therefore any region in which electric charges experience mechanical forces is called an electric field.

ELECTRIC FLUX

An electric field is represented by electric lines of force or electric flux. Electric flux emanates from a positive charged body as shown in Fig. 79.

Electric flux is equal to the electric charge Q coulombs.

$$Q = \text{electric flux (coulombs)}.$$

ELECTRIC FLUX DENSITY

The surface area of a sphere is given by $4\pi r^2$ where r is the radius of the sphere. The electric flux, Q, emanates in three dimensions from the sphere has an electric flux density

$$\boxed{D = \frac{Q}{A}} \quad \dots (2)$$

D stands for a displacement of electric flux density (coulombs per square metre) (C/m^2)
Q denotes electric flux in coulombs (C)
A denotes the surface area in square metres (m^2).

ELECTRIC FIELD STRENGTH

The electric field strength is given in newtons per coulomb or volts per metre (V/m) and is denoted by the symbol E.

If the potential difference between the two parallel plates is V and the distance separated by the plates is d, then the electric field strength is given by

$$\boxed{E = \frac{V}{d}} \quad \dots (3)$$

If a very small, positive point charge Q is placed at any point in an electric field and it experiences a force F, then the field strength E at that point is defined by the equation

$$\boxed{E = \frac{F}{Q}} \dots (4)$$

the electric field strength is the force in newtons per unit charge and its direction is that of the force F, thus the field strength E is a vector.

45

b. Expresses field strength in terms of potential gradient.

Potential gradient is the drop in potential per metre in the direction of the electric field.

The work done by a positive point charge of one coulomb in moving a short distance dx metres in the direction of the field is $E\,dx$ joules.

The potential difference in volts between two points is equal to the work done, in joules, in moving one coulomb of charge from one point to the other.

$$dV = -E\,dx$$

the potential gradient is therefore given by E as

$$\boxed{E = -\frac{dV}{dx}} \quad \ldots (5)$$

c. Establishes the relationship between electric field strength and electric flux density and defines relative permittivity and permittivity of free space.

The ratio of the electric flux density to the electric field strength gives the permittivity of the medium

$$\boxed{\in = \frac{D}{E}} \quad \ldots (6)$$

D = electric flux density (coulombs per unit area)
E = electric field strength (volts per metre)
$\in$ = absolute permittivity (farads per metre)

$\in = \in_{o}\in_{r}$ where $\in_{o}$ = permittivity of free space

$$\in_{o} = 8.85 \times 10^{-12} \text{ F/m}$$

$\in_{r}$ = relative permittivity (no dimensions)

$$\boxed{\in_{r} = \frac{\in}{\in_{o}}}$$

d. Defines capacitance as the constant of proportionality between charge and potential difference and establishes the relationship between capacitance and the physical dimensions of a pair of parallel plates.

THE CAPACITOR

Any two conductors between which an electric field can be maintained from a capacitor.

46

Consider two parallel plates as shown in Fig. 81 subject to a p.d. of V volts, one plate is positively charged and the opposite plate is negatively charged.

Fig. 81 Two parallel plate capacitor.

CAPACITANCE OF A PARALLEL PLATE CAPACITOR

The capacitance, $C = \dfrac{Q}{V}$ where Q is the charge in coulomb and V is the p.d. across the plates.

$$D = \frac{Q}{A}$$

$$E = \frac{V}{d}$$

$$\frac{D}{E} = \epsilon_o \epsilon_r$$

$$C = \frac{Q}{V} = \frac{DA}{Ed} = \frac{\epsilon_o \epsilon_r A}{d}$$

$$\boxed{C = \epsilon_o \epsilon_r \frac{A}{d}} \quad \text{... (8)}$$

For n plates of metal foil interleaved with a dielectric of mica or waxed paper, C is given by

$$\boxed{C = \frac{\epsilon_o \epsilon_r A(n - 1)}{d}} \quad \text{... (9)}$$

47

Fig. 82 Several parallel plate capacitor.

e. Derives expressions for energy stored by a capacitor

A p.d. of V volts is applied across a capacitor C, the charge on the capacitor is given as

$$Q = CV.$$

The charge Q is directly proportional to the p.d. applied, V

$$Q \propto V$$

The energy stored, W is equal to QV joules

$$W = QV.$$

Assuming that the capacitor is initially uncharged, the energy at that instant is zero. When V is applied, the energy stored is $W = QV$. The total energy stored in the capacitor is the average or mean value

$$W = \frac{0 + QV}{2} = \frac{1}{2} QV$$

$$W = \frac{1}{2} QV$$

but $Q = CV$

$$W = \frac{1}{2} (CV) V = \frac{1}{2} CV^2$$

$$W = \frac{1}{2} CV^2 \quad \dots (10)$$

the energy stored in a capacitor is thus proportional to the p.d. squared, therefore it is greatly dependent on the voltage.

f. Solves problems relating to uniform fields in simple dielectrics involving the relationships established in a to e.

WORKED EXAMPLE 14

A 150 μF capacitor is connected across a 150 V d.c. supply. Determine (i) the charge (ii) the energy stored.

SOLUTION 14

(i) $Q = CV = 150 \times 10^{-6} \times 150 = 22.5$ mC

(ii) $W = \dfrac{1}{2} CV^2 = \dfrac{1}{2} 150 \times 10^{-6} \times 150^2 = 1.69$ J.

WORKED EXAMPLE 15

A p.d. of 100 V d.c. is applied across the plates of a two-plate capacitor of capacitance 0.01 μF. The effective area of each plate is 100 mm^2 and the absolute permittivity of the dielectric is 150 pF/m.

Determine (i)   the electric flux density
           (ii)   the electric field intensity
           (iii) the relative permittivity

$\in_o = 8.85 \times 10^{-12}$ F/m.

SOLUTION 15

(i) $D = \dfrac{Q}{A} = \dfrac{CV}{A} = \dfrac{0.01 \times 10^{-6} \times 100}{100 \times 10^{-6}} = 0.01$ C/m^2

(ii) $E = \dfrac{D}{\in} = \dfrac{0.01}{150 \times 10^{-12}} = \dfrac{10^{10}}{150} = 6.67 \times 10^7$ V/m

(iii) $\in = \in_r \in_o$

$\in_r = \dfrac{\in}{\in_o} = \dfrac{150 \times 10^{-12}}{8.85 \times 10^{-12}} = 16.95.$

WORKED EXAMPLE 16

A two plate capacitor has an effective area of 75 cm^2 and the dielectric is air of 7.5 mm. A d.c. p.d. of 100 V is applied.

Calculate: (i)   the potential gradient in the dielectric
           (ii) the capacitance
           (iii) the electric flux
           (iv) the electric flux density
                 $\in_o$ = 8.85 × 10⁻¹² F/m.

SOLUTION 16

(i) $E = \dfrac{V}{d} = \dfrac{100}{7.5 \times 10^{-3}}$ = 13.3 KV/m

(ii) $C = \dfrac{\in_o \in_r A}{d} = \dfrac{8.85 \times 10^{-12} \times 1 \times 75 \times 10^{-4}}{7.5 \times 10^{-3}}$ = 8.85 pF

(iii) $Q = CV = 8.85 \times 10^{-12} \times 100 = 8.85 \times 10^{-10}$ C

(iv) $D = \dfrac{Q}{A} = \dfrac{8.85 \times 10^{-10}}{75 \times 10^{-4}} = 0.118 \times 10^{-6}$ C/m².

WORKED EXAMPLE 17

A capacitor is formed by two parallel metal plates each of area 1000 mm² and separated by a dielectric 0.5 mm thick. The capacitance of this arrangement is 500 pF and a d.c. p.d. of 20000 V is applied to the terminals.

Calculate: (i)   the charge on the plates
           (ii) the relative permittivity of the dielectric
           (iii) the electric flux density
           (iv) the energy stored
                 $\in_o$ = 8.85 × 10⁻¹² F/m.

SOLUTION 17

(i) $Q = CV = 500 \times 10^{-12} \times 20000 = 10\ \mu C$

(ii) $\in = \dfrac{D}{E} = \dfrac{\dfrac{Q}{A}}{\dfrac{V}{d}} = \dfrac{\dfrac{10 \times 10^{6}}{1000 \times 10^{-6}}}{\dfrac{20000}{0.5 \times 10^{-3}}} = \dfrac{0.01}{4 \times 10^{7}}$

      $\in = 0.25 \times 10^{-9}$ F/m

      $\in_r = \dfrac{\in}{\in_o} = \dfrac{0.25 \times 10^{-9}}{8.85 \times 10^{-12}} = 28.2$

(iii) $D = \dfrac{Q}{A} = \dfrac{10 \times 10^{-6}}{1000 \times 10^{-6}} = 0.01 \ C/m^2$

(iv) $W = \dfrac{1}{2} CV^2 = \dfrac{1}{2} \times 500 \times 10^{-12} (20000)^2 = 0.1 \ J$

g. Deduces expressions for the equivalent capacitance of capacitors connected in series and parallel, solves simple problems and compares predicted and measured values for series and parallel capacitor combinations.

CAPACITORS IN SERIES

Fig. 83 Capacitors in series

Let V_1 and V_2 be the two p.d.s across C_1 and C_2 respectively.

The total charge of the circuit is also the charge on each capacitor since $Q = It$ and I is the same when the capacitors are charged.

Used KVL $V = V_1 + V_2$ ... (1)

and since $Q = CV$ or $V = \dfrac{Q}{C}$, $V_1 = \dfrac{Q}{C_1}$ and $V_2 = \dfrac{Q}{C_2}$

and substituting in (1), $\dfrac{Q}{C} = \dfrac{Q}{C_1} + \dfrac{Q}{C_2}$

dividing each term by Q, we have

$$\boxed{\dfrac{1}{C} = \dfrac{1}{C_1} + \dfrac{1}{C_2}} \quad \text{... (2)}$$

If there are n capacitors in series, the formula is extended to

$$\boxed{\dfrac{1}{C} = \dfrac{1}{C_1} + \dfrac{1}{C_2} + \dfrac{1}{C_3} + \dots + \dfrac{1}{C_n}} \quad \text{... (3)}$$

51

CAPACITORS IN PARALLEL

The total charge of the system is shared between the two capacitors

$$Q = Q_1 + Q_2 \quad \dots (4)$$

Fig. 84 shows the circuit of two capacitors in parallel

Fig. 84 Two capacitors in parallel

$$V = \frac{Q_1}{C_1}, \; V = \frac{Q_2}{C_2}, \; V = \frac{Q}{C}.$$

Substituting these values in equation (4) we have

$$CV = C_1V + C_2V$$

dividing each term by V

$$\boxed{C = C_1 + C_2} \quad \dots (5)$$

The system can be extended to n capacitors in parallel, then the total capacitance

$$\boxed{C = C_1 + C_2 + C_3 + \dots + C_n} \quad \dots (6)$$

WORKED EXAMPLE 18

Determine the equivalent capacitance of the capacitor networks of Fig. 85, 86, 87, 88.

Fig. 85 Capacitors in series

Fig. 86 Capacitors in series and parallel

Fig. 87 Capacitors in parallel

Fig. 88 Capacitors in series and parallel

SOLUTION 18

(i)   Referring to Fig. 85

$$\frac{1}{C} = \frac{1}{C_1} + \frac{1}{C_2} + \frac{1}{C_3} = \left(\frac{1}{1} + \frac{1}{2} + \frac{1}{3} \right) 10^6 = \frac{11}{6} \times 10^6$$

therefore $C = \frac{6}{11} \, \mu F$

or $C = \dfrac{C_1 C_2 C_3}{C_1 C_2 + C_1 C_3 + C_2 C_3} = \dfrac{1 \times 2 \times 3}{1 \times 2 + 1 \times 3 + 2 \times 3} \times 10^{-6} = \dfrac{6}{11} \, \mu F$

53

(ii)   Referring to Fig. 86

$$C = C_{T_1} + C_{T_2} = \frac{C_1 C_2}{C_1 + C_2} + \frac{C_3 C_4}{C_3 + C_4} = \left(\frac{1 \times 2}{1 + 2} + \frac{3 \times 4}{3 + 4} \right) \mu F$$

$$= \left(\frac{2}{3} + \frac{12}{7} \right) \mu F = \frac{50}{21} \mu F$$

(iii) Referring to Fig. 87

$$C = C_1 + C_2 + C_3 = (10 + 20 + 30) \mu F = 60 \mu F$$

(iv)   Referring to Fig. 88

$$C_{T_1} = \frac{C_1 C_2}{C_1 + C_2} = \frac{1 \times 5}{1 + 5} \mu F = \frac{5}{6} \mu F$$

$$C_{T_2} = C_3 + C_4 = (10 + 10) \mu F = 20 \mu F$$

$$C = \frac{C_{T_1} C_{T_2}}{C_{T_1} + C_{T_2}} = \frac{\left[\dfrac{5}{6} \right] 20}{\dfrac{5}{6} + 20} \mu F = 0.8 \mu F$$

$$C_T = (0.8 + 100) \mu F = 100.8 \mu F.$$

WORKED EXAMPLE 19

For the circuit shown in Fig. 89 calculate:
(i)   the equivalent capacitance
(ii)   the p.d. across the 30 μF
(iii) the total energy stored.

Fig. 89   Capacitors in series and parallel

54

SOLUTION 19

(i) The equivalent capacitance can be found by first taking C_1 and C_2 in parallel

$C_{T_1} = C_1 + C_2$, then taking the two capacitors C_{T_1} and C_3 in series,

$$\frac{C_{T_1} C_3}{C_{T_1} + C_3} = C_T, \text{ then taking } C_T \text{ and } C_4 \text{ in parallel, } C_T + C_4 = C_{T_2} \text{ and}$$

finally $C = \dfrac{C_5 C_{T_2}}{C_5 + C_{T_2}}$, the equivalent capacitance

$$C_{T_1} = C_1 + C_2 = (15 + 10)\ \mu F = 25\ \mu F$$

$$C_T = \frac{C_{T_1} C_3}{C_{T_1} + C_3} = \frac{25 \times 15}{25 + 15}\ \mu F = 9.375\ \mu F$$

$$C_{T_2} = C_T + C_4 = (9.375 + 10)\ \mu F = 19.375\ \mu F$$

$$C = \frac{C_5 C_{T_2}}{C_5 + C_{T_2}} = \frac{30 \times 19.375}{30 + 19.375}\ \mu F = 11.77\ \mu F.$$

(ii) In order to find the p.d. across 30 μF capacitor, we have to consider Fig. 90 where C_5 and C_{T_2} are in series across the 100 V d.c. supply

Fig. 90 Equivalent circuit of capacitors

The total charge of the system $Q = CV = 11.77 \times 10^{-6} \times 100$

$$Q = 1177\ \mu C$$

The p.d. across 30 μF is found from $V = \dfrac{Q}{C_5}$

$$V = \frac{1.177 \times 10^{-3}}{30 \times 10^{-6}} = 39.2 \text{ volts.}$$

55

(iii) $W = \frac{1}{2} CV^2 = \frac{1}{2} \times 11.77 \times 10^{-6} \times 100^2 = 0.059$ J

$W = 59$ mJ

h. Identifies and distinguishes between capacitors of differing construction and characteristics and relates dielectric strength to working voltage.

Most practical capacitors have their plates separated by an insulating material which is called a dielectric. It was shown earlier that the effect of introducing a dielectric was to increase the capacitance of the capacitor, according to the formula

$$C = \frac{\epsilon_o \epsilon_r A}{d}$$

where ϵ_r is the relative permittivity of the dielectric constant.

It is important that the dielectric shall be able to withstand the potential gradient applied on it when the capacitor is charged.

The resistance of the dielectric layer is known as *the insulation resistance* which must be extremely high for a good component, in the order GΩ. The insulation resistance depends on the area and thickness of the dielectric and falls with increase in temperature.

For a large capacitance, the thickness of a dielectric is thin and the insulation resistance therefore will be low.

If the potential gradient applied to a capacitor is exceeded from the quoted value by the manufacturer, the insulation is likely to break down.

A capacitor is rated to withstand a maximum *working voltage.*

Different dielectric materials differ in the potential gradient they can withstand and to test a particular material, a dielectric of 1 mm thickness thick is placed between the two electrodes, and the p.d. is then increased gradually across the plates until the insulation resistance breaks down. The potential gradient sufficient to cause breakdown of the insulation is called the *dielectric strength* of the material and is expressed in KV per mm. The dielectric strength KV/mm for some materials is given.

Material	Dielectric Strength KV/mm
Glass	20
Mica	40 to 100
Polythene	40
Barium Titanate	4
Air	1
Oil (Insulating)	40

SUMMARY 3

$$F = \frac{Q_1 Q_2}{4\pi \in r^2} \ (\text{N})$$

$$\in \ = \ \in_r \in_o \ (\text{F/m})$$

$$D = \frac{Q}{A} \ (\text{C/m}^2)$$

$$C = \frac{\in_o \in_r A}{d} \ (\text{F})$$

$$E = \frac{V}{d} \ (\text{V/m})$$

$$E = \frac{F}{Q} \ (\text{N/C})$$

$$C = \frac{\in_o \in_r A (n-1)}{d} \ (\text{F})$$

$$E = -\frac{dV}{dx} \ (\text{V/m})$$

$$Q = CV \ (\text{C})$$

$$\in \ = \frac{D}{E} \ (\text{F/m})$$

$$W = \frac{1}{2} CV^2 \ (\text{J})$$

Capacitors in series $\dfrac{1}{C} = \dfrac{1}{C_1} + \dfrac{1}{C_2} + \ldots + \dfrac{1}{C_n}$

Capacitors in parallel $C = C_1 + C_2 + \ldots + C_n$.

EXERCISES 3

1. A parallel plate capacitance of 100 pF. What will be the new value of capacitance if the effective plate area is reduced by 50% and the dielectric thickness is doubled?

   (Ans. 25 pF).

2. (a) What factors determine the capacitance existing between a parallel metal plates?

   (b) Write down the formula for the capacitance connecting the above factors and state the unit of each physical quantity.

   (c) For a given capacitor write down the expressions starting the units employed relating:

   (i) the potential difference and charge

   (ii) the potential difference and stored energy.

3. Two air-spaced metal plates insulated from each other are mounted so as to form a parallel plate capacitor. When the distance between the plates is 2.5 mm the capacitance is 5 pF. Determine the capacitance when the spacing is increased to 3.5 mm.

   How will the value of the original capacitor be affected if the whole is immersed in oil having a relative permittivity of 5?

   (Ans. 3.57 pF, 25 pF)

4. A p.d. of 15 KV is applied to the terminals of a capacitor consisting of two circular plates, each having an area of 100 cm^2, separated by a dielectric 2.5 mm thick. If the capacitance is 100 pF, calculate:

   (i) the electric flux

   (ii) the electric flux density

   (iii) the relative permittivity of the dielectric $\left(\in_o = 8.85 \times 10^{-12} \text{ F/m} \right)$.

   (Ans. (i) 1.5 μC (ii) 1.5 $\times$ 10^{-4} C/m^2 (iii) 2.83)

5. A capacitor is charged with 100 μC, if the energy stored is 1 μJ find

   (i) the p.d.

   (ii) the value of the capacitance.

   (Ans. (i) 0.02 V (ii) 5000 μF)

6. Two capacitors A and B have capacitances of 10 μF and 20 μF respectively. Calculate the voltage across A and the energy stored in A when A and B are connected (i)   in series and

    (ii)   in parallel with a d.c. supply of 20 V in both cases.

(Ans. (i) 13.3 V, 888 μJ   (ii) 20 V, 2 mJ)

7. Two capacitors 10 μF and 30 μF are connected in series. Across both of these is connected a capacitor of 20 μF. The whole combination is connected across a 100 V d.c. supply.

    Calculate: (i)   the equivalent capacitance

    (ii)   the p.d. across the 30 μF capacitor

    (iii)   the charge on 10 μF capacitor

    (iv)   the energy stored by the 20 μF capacitor.

(Ans. (i) 27.5 μF   (ii) 25 V   (iii) 750 μC   (iv) 0.1 J)

8. A 1 μF capacitor charged to a p.d. of 100 V is connected across a second uncharged capacitor of 2 μF, the source of voltage being removed before the second capacitor is added to the first.

    Find (i)   the resulting p.d. across the combined capacitors

    (ii)   the energy stored in each capacitor before and after paralleling. Comment on the results obtained.

    (iii)   if the two capacitors were discharged and then connected in series across the 100 V supply what would be the voltage across each capacitor?

(Ans.   (i) 33.3 V   (ii) 0.005 J, 0 J; 554 mJ, 1109 μJ, 5 mJ, 1.66 mJ, Loss 3.34 mJ   (iii) 66.7 V, 33.3V)

9. Calculate: (i)   the charge on each capacitor before C is connected

    (ii)   the total energy stored

    (iii)   the value of C which, when connected in parallel with the 20 μF capacitor, would change V_2, to 60 V with the same supply of Fig. 91 (next page).

Fig. 91 Series-parallel capacitors sharing charge.

(Ans. (i) 1.33 mC (ii) 0.132 J, 140 V, 60 V (iii) 3.3 μF)

10. Determine the total value of capacitance for each capacitor system in Fig. 92 and Fig. 93.

Fig. 92 Series-parallel capacitors

Fig. 93 Series-parallel capacitors

When C is (a) 1 μF and C is (b) 10 μF.

(Ans. (a) 0.5 μF, $\frac{3}{10}$ μF, (b) 5 μF , 3 μF)

4. MAGNETIC FIELDS

4. Applies the fundamental laws governing magnetic fields to the solution of problems relating to magnetic circuits and materials.

a. Introduces the concept of the magnetic field and magnetic flux to explain the forces of attraction and repulsion between magnetised bodies and defines magnetic field strength.

MAGNETIC FIELD

A number of magnetic lines make up a magnetic field. A bar magnet or a permanent magnet possesses a magnetic field which can be detected by a small compass needle or by sprinkling iron filings. The field due to a bar magnet is illustrated in Fig. 94, while the magnetic field due to a solenoid is shown in Fig. 95

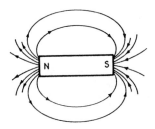

Fig. 94 The magnetic field round a bar magnet

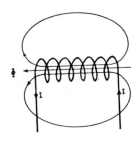

Fig. 95 The magnetic field due to a solenoid

MAGNETIC FLUX

The magnetic lines shown are flowing from the north pole, N into the south poles. The magnetic lines *do not* intersect.

61

The magnetic lines of force or the magnetic flux is denoted by the Greek letter ϕ and it is expressed in webers (Wb). Faraday introduced the lines of flux as a pictorial method of representing the distribution of a magnetic field and they are used to define field quantitatively.

The magnetic fields of two like poles and two unlike poles are shown in Fig. 96 and Fig. 97.

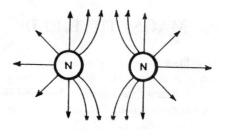

Fig. 96 Like magnetic poles

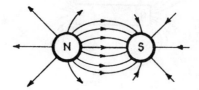

Fig. 97 Unlike magnetic poles

In Fig. 96 the magnetic field between the like poles is weak, and in Fig. 97 the magnetic field between the unlike poles is strong. The former field exhibits a force of repulsion and the latter field exhibits a force of attraction.

MAGNETIC FIELD STRENGTH OR MAGNETISING FORCE

The magnetic potential difference or the magnetomotive force (m.m.f.) is produced by the current flow in a number of turns N. the m.m.f. is given by

$$F = IN \text{ (ampere-turns)} \quad \text{At}$$

The same current is used N times and it is shown in Fig. 98

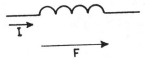

Fig. 98 The magnetic potential difference (m.m.f.)

This is analogous to the e.m.f. in an electric circuit.

The magnetic field strength, H, is the m.m.f., F, per metre length, l of the flux path

$$H = \frac{F}{l} = \frac{IN}{l} \text{ (At/m)}$$

where l is the length of the flux path or the magnetic length.

WORKED EXAMPLE 20

A toroid shown in Fig. 99 is wound uniformly with 1000 turns and carries a current $I = 2$ A, the mean diameter of the toroid is 30 cm. Determine the magnetic field strength.

Fig. 99 The toroid.

SOLUTION 20

The magnetic flux, ϕ, flows in the direction shown according to the corkscrew rule and the magnetic length is πD, the flux path

$$H = \frac{IN}{l} = \frac{2 \times 1000}{\pi 0.3} = 2122 \text{ At/m}$$

b. Establishes the relationship between magnetic field strength and magnetic flux density and defines relative permeability and the permeability of free space

MAGNETIC FLUX DENSITY

The magnetic line of flux ϕ webers which crosses an area A m^2 is termed as magnetic flux density, B

$$B = \frac{\phi}{A} \text{ (tesla or Wb/m}^2\text{)}$$

In Fig. 100, the magnetic flux, ϕ, crosses the cross sectional area, A, to give the flux density.

Fig. 100 Flux density

PERMEABILITY

The relationship between magnetic field strength and magnetic flux density is given by the equation

$$\frac{B}{H} = \mu$$

and $\mu = \mu_o \mu_r$.

The ratio of B to H is termed as absolute permeability μ where μ_o is the permeability of free space which is a constant given as 4π 10^{-7} H/m (henrys per metre) and μ_r is the relative permeability which is a number and has no dimensions ($\mu_r = 1$ for a non-magnetic material).

The permeability is a measure of the magnetic performance of a material.

c. Investigates the effect of the core material on the performance of an electromagnet, compares magnetisation characteristics of typical ferromagnetic materials obtained by measurement and deduces the range of values of relative permeability.

AN ELECTROMAGNET

A number of turns are wound on a ferromagnetic cores as shown in Fig. 101.

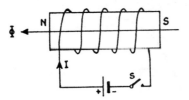

Fig. 101 The electromagnet.

When the switch, S, is closed, current I flows in the windings thus setting up a magnetic flux as shown by the letter ϕ. This is the principle of an electromagnet. Its advantages over the permanent bar magnet is that it may be switched on and off as required. It is a temporary magnet and must be made of soft iron.

Referring to the previous formulae.

$$B = \frac{\phi}{A}, \quad H = \frac{IN}{l} \quad \text{and} \quad \frac{B}{H} = \mu = \mu_o \mu_r.$$

If the core is non-magnetic then $\mu_r = 1$ and the ratio $\frac{B}{H}$ is minimum. If the core

is ferromagnetic then μ_r is large (e.g. 7000) and $\frac{B}{H}$ is large.

MAGNETIC MATERIALS

Iron, nickel, cobalt and certain alloys of these metals can be made into strong magnets, and are known as *ferromagnetic* materials.

Steel is an alloy of iron, made by adding small percentage of carbon to pure iron and it is a much harder metal than pure iron.

A permanent magnet can be made by placing a steel bar or rod inside a solenoid and the current switched on. The steel becomes a magnet and can pick up pieces of certain metals.

Alloys of nickel and cobalt are used for making powerful permanent magnets and are called ferrites. These magnetize strongly and are not easily demagnetized. Ferrites have a high electrical resistance, and hence the current and power loss are extremely small.

The graph of B against H for magnetic and non-magnetic material is shown in Fig. 102.

Fig. 102 $\frac{B}{H}$ graphs for magnetic and non-magnetic materials.

The magnetization curve for a magnetic material is non-linear and for a non-magnetic material is linear. The former has a variable permeability and the latter a constant permeability, this is shown in Fig. 103 where μ_r is plotted against H.

Fig. 103 $\frac{\mu_r}{H}$ graphs for magnetic and non-magnetic.

There are 'soft' magnetic materials and 'hard' magnetic materials. The former have high saturation B and low loss (iron, steel, silicon steel, nickel-iron alloys $(30 - 70\% \, \text{Ni})$, cobalt-iron alloys, manganese and aluminium oxides (ferrites) and the latter have high residual magnetic effect. The BH curves for the soft and hard magnetic materials are shown in Fig. 104.

Fig. 104 BH curves for soft and hard materials.

Magnetisation curves for silicon iron, mild steel, and cast iron are shown in Fig. 105. the relative permeability plotted against the flux density for these materials are shown in Fig. 106.

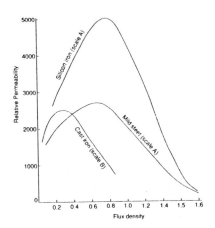

Fig. 105 *BH* curves for various materials

Fig. 106 Curves of relative permeability against flux density for various materials.

It can be deduced from the graphs of μ_r / B that values of μ_r = 5000 for silicon iron at B = 0.75 T, μ_r = 2500 for cast iron at B = 0.225 T and μ_r = 2600 for mild steel at B = 0.625 T may be obtained.

d. Relates magnetomotive force, reluctance and magnetic field strength and solves problems involving magnetic circuits having not more than a single change of dimension, material or air-gap using data from magnetisation curves.

RELUCTANCE

The student is familiar with the word resistance of an electric circuit, that is the controlling component of current in an electric circuit is the resistor.

Reluctance, S, is the analogous quantity in a magnetic circuit

$$\text{Reluctance} = S = \frac{F}{\phi} = \frac{\text{Magnetomotive force}}{\text{flux}} \quad \dots (1)$$

and it is expressed in A/Wb (amperes/weber). Reluctance is analogous to resistance in an electric circuit. The formula of equation (1) is analogous to the formula of resistance, $R = V/I$ which is ohm's law.

Fig. 107 which indicates the property of the reluctance which is the magnetic resistance of the magnetic circuit.

Fig. 107 Reluctance

$$S = \frac{F}{\phi} = \frac{F}{BA} = \frac{F}{\mu HA} . \quad \dots (2)$$

The relationship between m.m.f., reluctance and magnetic field strength, H is given

$$S = \frac{F}{\mu HA} = \frac{F}{\mu_o \mu_r HA} . \quad \dots (3)$$

WORKED EXAMPLE 21

A coil of 500 turns, wound on ring of magnetic material, carries a current of 1 A. The length and cross-sectional area of the flux path in the ring are 500 mm and 1000 mm² respectively. The relative permeability, $\mu_r = 3500$ and $\mu_o = 4\pi \times 10^{-7}$ H/m. Determine (i) the magnetic flux density
(ii) the reluctance.

SOLUTION 21

$$S = \frac{F}{\phi} = \frac{IN}{BA} \qquad \frac{B}{H} = \mu_o \mu_r = \frac{B}{IN/l} = \frac{Bl}{IN}$$

(i) $B = \mu_o \mu_r \dfrac{IN}{l} = \dfrac{4\pi \times 10^{-7} \times 3500 \times 1 \times 500}{0.5} = 4.4$ T

(ii) $S = \dfrac{IN}{BA} = \dfrac{1 \times 500}{4.4 \times 10^{-3}} = 113636$ A/Wb.

WORKED EXAMPLE 22

A piece of iron has a magnetic length of 150 mm a cross-sectional area of 500 mm², and a relative permeability of 550. Find the reluctance of the iron ($\mu_o = 4\pi \times 10^{-7}$ H/m).

68

SOLUTION 22

$$S = \frac{F}{\phi} = \frac{IN}{BA} = \frac{IN}{H\mu_o\mu_r A} = \frac{INl}{IN\mu_o\mu_r A} = \frac{l}{\mu_o\mu_r A}$$

$$S = \frac{l}{\mu_o\mu_r A} = \frac{0.15}{4\pi \times 10^{-7} \times 550 \times 500 \times 10^{-6}} = 434059 \text{ A/Wb.}$$

AIR GAP

A toroid of a magnetic material has a radial cut as shown in Fig. 108.

Fig. 108 A radical cut in a toroid.

The magnetic flux set up in the core is shown which also passes through the air gap.

The magnetic circuit is now made up of two reluctances in series since the magnetic flux is the same in the core and in the air gap, assuming no magnetic leakage occurs.

The magnetic lengths are l_i that of the iron, and l_a, that of the air gap.

The total reluctance of the magnetic circuit is given by

$$S = S_i + S_a = \frac{l_i}{\mu_o\mu_r A} + \frac{l_a}{\mu_o A}$$

since $\mu_r = 1$ for the air gap.

It is obvious that the effect of the air gap is to increase the reluctance of the magnetic circuit. This argument is analogous to that of one electric circuit, the total resistance is equal to the sum of the resistances since the same current flows through each resistor.

COMPARISON BETWEEN ELECTRIC AND MAGNETIC CIRCUIT

ELECTRIC CIRCUITS			MAGNETIC CIRCUITS		
TERM	SYMBOL	UNIT	TERM	SYMBOL	UNIT
Current	I	A	Flux	ϕ	Wb
Electromotive force	E	V	Magnetomotive force	F	A
Current Density	J	A/m^2	Flux density	B	Wb/m^2 (T)
Resistance	R	Ω	Reluctance	S	A/Wb
Conductivity	σ	S/m	Permeability	μ	H/m
Resistivity	ϱ	Ωm	Reluctivity	$\dfrac{1}{\mu}$	m/H

PERMEABILITY IS ANALOGOUS TO CONDUCTIVITY

$$R = \varrho \frac{l}{A} \text{ or } \varrho = \frac{RA}{l} \quad \text{or} \quad \sigma = \frac{1}{\varrho} = \frac{l}{RA} = \text{conductivity}$$

$$\mu = \frac{B}{H} = \frac{\phi/A}{F/l} = \frac{\phi l}{FA} = \frac{l}{SA} \text{ permeability}$$

$$S = \frac{l}{\mu A} = \left(\frac{1}{\mu}\right)\frac{l}{A} \text{ analogous to } R = (\varrho)\frac{l}{A}$$

therefore $\dfrac{1}{\mu}$ is analogous to ϱ or μ is analogous to σ.

WORKED EXAMPLE 23

A 1500 turn coil is wound on an iron ring having a c.s.a. of 10 cm^2, and a mean diameter of 0.15 m. There is radial saw cut 1 mm wide in the ring.

Calculate the current in the coil for a flux of 1 mWb in the ring. The $\dfrac{B}{H}$ curve for the iron is as follows:

B (T)	0.62	0.88	1.03	1.13	1.18
H (A/m)	150	250	350	450	550

$\mu_o = 4\pi \times 10^{-7}$ H/m

SOLUTION 23

The magnetisation curve $\dfrac{B}{H}$ is plotted on Fig. 109.

Fig. 109 $\dfrac{B}{H}$ Curve

$$B = \frac{\phi}{A} = \frac{1 \times 10^{-3}}{10 \times 10^{-4}} = 1 \text{ T.}$$

From the graph of Fig. 109, $H_i = 320$ A/m corresponding to $B = 1$ T. If the air gap is not taken into account then $H_i = 320 = \dfrac{I_i N}{0.471}$

$$l_i = \pi D = \pi \times 0.15 = 0.471 \text{ m}$$

$$H_i = \frac{I_i 1500}{0.471} \text{ or } 1500 \, I_i = 0.471 \times 320$$

71

current due to the iron, $I_i = \dfrac{320 \times 0.471}{1500} = 0.1$ A

$$H_a = \frac{S_a\phi}{l_a} = \frac{l_a\phi}{\mu_o A l_a} = \frac{\phi}{\mu_o A} = \frac{1 \times 10^{-3}}{4\pi \times 10^{-7} \times 10 \times 10^{-4}} = 795775 \text{ A/m}.$$

The total magnetic field intensity

$$H = 320 + 795775 = 796095 \text{ A/m}.$$

Current due to the air gap

$$I_a = \frac{H_a l_a}{N} = \frac{795775 \times 0.001}{1500} = 0.531 \text{ A}.$$

The total current $= 0.1 + 0.531 = 0.631$ A.

e. *Displays hysteresis loops on a C.R.O. and observes the effects of variation of magnetic material and magnetic field strength.*

HYSTERESIS

Hysteresis is a Greek word meaning 'lagging' and in this context, means that the flux density, B, is lagging behind the magnetic field intensity or magnetising force, H, when a specimen of ferromagnetic material is taken through a cycle of magnetisation.

The specimen is initially in a demagnetised state. As H increases, the magnetisation curve is non-linear as we say previously and is shown as OA in Fig. 110.

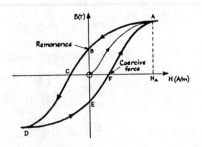

Fig. 110 Hysteresis loop

72

At A it saturates, that is, B is substantially constant as H increases beyond the point A. If H or I is decreased from its value H_A, the curve now follows the path AB, where at B, $OB = B_r$ when $H = OB_r$ = remanent flux density. Reversing the current, the path follows, BC, CD, DE, EF and one complete cycle is traced when it ends at A.

$OF = H_c$ = coersive force, when B = zero tesla. The closed loop $ABCDEFA$, is called a hysteresis loop when H has been reduced from H_A to zero the iron is still magnetised in the positive direction and the magnetic flux density at that point is called the *remanent flux density*, since there is a residual magnetism in the iron. The amount of negative magnetising force which must be applied in order to annul the remanent flux density is called the *coersive force*.

The remanent flux density corresponding to saturation of the material is called *remanence* and the corresponding coercive force is called coercivity.

f. States that hysteresis loop is proportional to the area of the loop and explains the importance of hysteresis loss for a.c. excited devices.

The energy stored in the magnetic material is proportional to the area of the loop. It can be shown that the energy expended per cycle per cubic metre is equal to the area of the hysteresis loop in joules.

Steinmitz devised an empirical formula, namely

Energy dissipated $= \eta B^{1.6}$ joules per m^3 per cycle

where η is the hysteresis coefficient and has numerical value between 250 and 300 for 4% silicon steel, 500 for sheet steel, 750 - 3000 for cast iron steel and 3000 - 4000 for cast iron.

Energy dissipated per second $= \eta f B^{1.6}$ in joules per m^3 where f is the frequency. Therefore the power loss $= \eta f B_m^{1.6}$ W/m^3.

No field.   Weak field.

Stronger field.

Fig. 111

To produce the extra flux density we require an intrinsic magnetising force H_i such that $B' = \mu_o H_i$ and $B = \mu_o(H + H_i)$

H_i is called the <u>intensity of magnetisation</u> of the core.

Now $B = \mu H = \mu_r \mu_o H = \mu_o(H + H_i)$ and $\mu_r = \dfrac{H + H_i}{H} = 1 + \dfrac{H_i}{H} = 1 + \chi$

where χ (chi) is the <u>magnetic susceptibility</u> of the core material.

MAGNETIC CLASSIFICATION OF MATRIALS

Although it is common to divide materials into magnetic and non-magnetic, experiment shows that most materials exhibit some magnetic properties if the magnetising force is great enough.

A FERROMAGNETIC MATERIALS
(e.g. iron, steel, nickel, cobalt and many alloys)

Here opposite polarity is induced in the end nearer the inducing pole.

A rod of the material sets itself in line with the field.

μ_r is not constant and may be large. Thus B is much greater in the material than it would be in air. μ_r for iron is 7000 max. For permalloy $\mu_r = 100000$.

B PARAMAGNETIC MATERIALS
(e.g. aluminium, manganese, copper, platinum)

As for ferromagnetic materials opposite polarity is induced and a rod sets itself in line with the field.

χ is positive and μ_r is greater than 1. Thus the flux density is slightly greater in the material than it would be in air. χ is constant and independent of H.

For aluminium $\chi = 2.2 \times 10^{-5}$. $(\mu_r = 1.000022)$

C DIAMAGNETIC MATERIALS

(e.g. antimony, bismuth, quartz)

In this case the same polarity is induced in the end nearer the inducing pole. A rod of the material sets itself across the magnetic field.

χ is constant but negative and μ_r is smaller than 1. Thus the flux density is slightly less in the material than it would be in air.

For bismuth $\chi = -2 \times 10^{-4}$ ($\mu_r = 0.9998$).

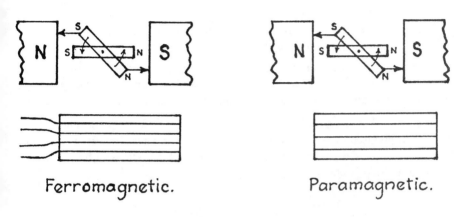

Ferromagnetic.   Paramagnetic.

Fig. 112

THE MAGNETISATION CURVE OR B-H CURVE OF A FERROMAGNETIC MATERIAL

The curve must be obtained experimentally.

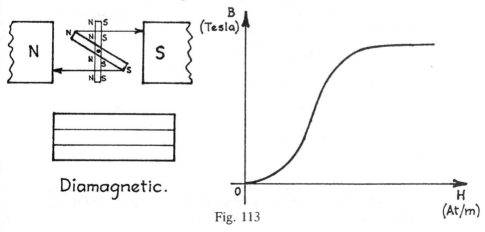

Diamagnetic.

Fig. 113

75

As the unmagnetised material is subjected to increasing magnetising force the intensity of magnetisation increases until it reaches a maximum value. The material is then saturated and any increase dH produces an increase dB = μ_o(dH) i.e. μ_r = 1. In the saturated state the magnetic domains are completely aligned with the field.

HYSTERESIS

If after saturation has been reached the magnetising force is reduced to zero the magnetic state of the material lags behind the magnetisation state and the descending part of the curve lies above the ascending part.

This is known as the hysteresis effect.

When H becomes zero the material still retains some magnetism. This is to the domains tending to stay aligned with the field.

In order to reduce B to zero it is necessary to apply a magnetising force in the reverse direction. The value of H needed to reduce B to zero is the coercive force H_c. This is a measure of the tenacity with which the material retains its magnetism.

If the magnetising force is further increased in the reverse direction the material becomes magnetised in the reverse direction and reaches saturation. If H is reduced to zero and increased in the original direction a similar hysteresis effect is shown and a closed loop is obtained.

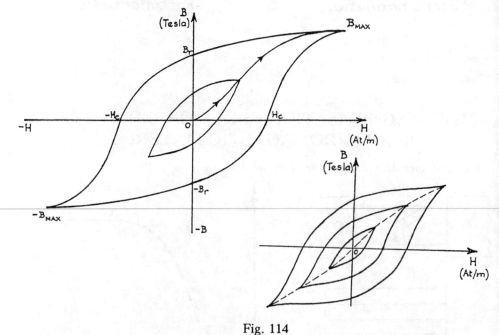

Fig. 114

The hysteresis loop shown assumes that saturation is reached. However, even if saturation is not reached a hysteresis loop is still obtained but this will lie inside the previous loop.

76

Once a piece of ferromagnetic material has been magnetised its magnetisation curve never passes through the origin again. The only way to completely demagnetise the material again is to carry it round a series of hysteresis loops of diminishing size. This can be done by passing an alternating current through the solenoid producing the magnetising force. If the current is gradually reduced to zero the material is left in an unmagnetised state.

HYSTERESIS LOSS

Work has to be done in rotating the domains both in magnetisation and demagnetisation. This work appears as heat. It can be shown that the hysteresis loss per cycle is proportional to the area of the hysteresis loop.

The diagram shows typical hysteresis loops for soft iron and tungsten steel. Although the soft iron has a greater remanence than the steel it is not as suitable for use as a permanent magnet since its coercivity is much less. This means that soft iron easily loses its magnetism as a result of knocks. Many alternating current machines subject magnetic material to alternating magnetising forces which means that the material performs hysteresis loops with the corresponding hysteresis loss. Soft iron would be more useful than tungsten steel since the area of its hysteresis loop is smaller than that of the tungsten steel.

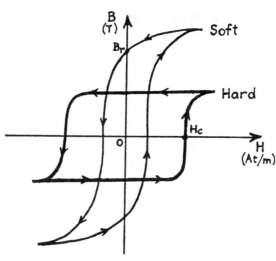

Fig. 115

THE MAGNETIC CIRCUIT

The term <u>electric circuit</u> refers to the particular path in which the electric current is constrained to flow.

The term magnetic circuit refers to the path taken by the magnetic flux, but it is not as well defined as the electric circuit.

Big differences:-

(i)   there is no flow in a magnetic circuit.

(ii) energy is needed to maintain an electric current, but no energy is needed to maintain a magnetic flux once it is set up

(iii) the difference between the resistivity of a good conductor and air is of the order of 10^{20}. The maximum valve of μ_r is only a few thousand. Thus an electric current is effectively confined to the conductors whereas there is considerable leakage of the magnetic field. This will be ignored.

MAGNETO-MOTIVE FORCE AND RELUCTANCE

In the electric circuit the fundamental quantities are e.m.f., current and resistance. In the magnetic circuit they are m.m.f., flux and reluctance. E.m.f. and m.m.f. correspond in that they are necessary for the production of the current and the flux. Current and flux correspond in that each gives the circuit its characteristic effects. Resistance and reluctance correspond in that both convey the idea of opposition.

For resistance we have
$$\text{Resistance} = \text{resistivity} \times \frac{\text{length}}{\text{area}}.$$

For reluctance we have
$$\text{Reluctance} = \text{reluctivity} \times \frac{\text{length}}{\text{area}}.$$

However, it is more usual to use the permeability which is the reciprocal of the reluctivity.

Thus $R = \varrho \dfrac{l}{A}$ and $S = \dfrac{1}{\mu} \cdot \dfrac{l}{A} = \dfrac{1}{\mu_r \mu_o} \cdot \dfrac{l}{A}$.

The flux density inside a solenoid is effectively constant and is given by

$$B = \mu_r \mu_o H = \mu_r \mu_o \frac{NI}{l}.$$

If the area of the field within the solenoid is A the flux $\phi = BA$

i.e. $\phi = \mu_r \mu_o \dfrac{NI}{l} A$

Rewriting $NI = \phi \times \dfrac{1}{\mu_r \mu_o} \cdot \dfrac{l}{A}$ (i.e. $NI = \text{flux} \times \text{reluctance}$)

Comparing this with the equation e.m.f. = current $\times$ resistance we see that NI is analagous to e.m.f. and is the m.m.f. of the circuit.

Summarising m.m.f. $= NI$; $S = \dfrac{1}{\mu_r \mu_o} \cdot \dfrac{l}{A}$ m.m.f. $= \phi \times S$

The unit of S is At/Wb,

RELUCTANCES IN SERIES

When different parts of a magnetic circuit have different dimensions or are made of different materials, then

78

(i)   the same flux is assumed to span each portion;

(ii)   the total reluctance equals the sum of the separate reluctances.

$$\text{i.e. } NI = \phi(S_1 + S_2 + S_3 + \text{etc.})$$

Example   A metal ring is 250 mm in mean diameter and has a cross-section of 500 mm^2. It is wound with a coil having 1000 turns. The ring consists of two semicircular pieces of cast iron and cast steel respectively.

Calculate the current required to produce a flux of 0.30 mWb in the ring. μ_r for cast iron = 175; μ_r for cast steel = 1500.

Supposing two pieces of brass each 2.0 mm thick are inserted between the semicircular pieces of iron and steel, what current is now needed to produce a flux of 0.30 mWb in the ring? μ_r for brass = 1.

For cast iron $S = \dfrac{1}{\mu_o \times 175} \cdot \dfrac{\pi \times 0.125}{500 \times 10^{-6}} = 3.57 \times 10^6 \text{At/Wb.}$

For cast steel $S = \dfrac{1}{\mu_o \times 1500} \cdot \dfrac{\pi \times 0.125}{500 \times 10^{-6}} = 0.42 \times 10^6 \text{At/Wb.}$

Total reluctance = 3.99×10^6 At/Wb.

$\therefore 1000\, I = 0.30 \times 10^{-3} \times 3.99 \times 10^6$

$I = 0.3 \times 3.99 = 1.2$ A.

For the brass inserts $S = \dfrac{1}{\mu_o} \cdot \dfrac{4 \times 10^{-3}}{500 \times 10^{-6}} = \dfrac{10^{-3} \times 10^7}{\pi \times 500 \times 10^{-6}}$

$S = 6.4 \times 10^6$ At/Wb.

Total $S = 10.4 \times 10^6$ At/Wb

$1000\, I = 0.3 \times 10.4 \times 10^6 \times 10^{-3}$

$I = 3.1$ A.

SUMMARY 4

$F = IN$ $S = \dfrac{F}{\phi}$

$H = \dfrac{IN}{l}$ $S = \dfrac{l}{\mu_o \mu_r A}$

$B = \dfrac{\phi}{A}$ $S = S_i + S_g = \dfrac{l_i}{\mu_o \mu_r A} + \dfrac{l_g}{\mu_o A}$

$\dfrac{B}{H} = M$ for series magnetic circuits $\mu = \mu_o \mu_r$

EXERCISES 4

1. A ring specimen of cast steel has a mean length of 50 cm and a cross sectional area of 10 cm².

    If 300 turns of wire are wound uniformly round the ring, calculate the current required in these turns to set up a magnetic flux of 750 μ Wb in ring and gap. (μ_r = 1000 for cast steel and μ_o = $4\pi \times 10^{-7}$ H/m. (Ans. 1 A)

2. A perspex toroid has a mean circumference of 75 cm and a c.s.a. of 10 cm². The coil consists of 500 turns and carries a current of 5 A.

    Calculate: (i)   the magnetomotive force
                (ii) the magnetic field intensity
                (iii) the flux density
                (iv) the total magnetic flux

    (Ans. (i) 2500 A (ii) 3333 A/m (iii) 4.19 m T (iv) 4.19 μ Wb)

3. A coil having 1000 turns is wound on an iron core having a mean length of 100 cm and c.s.a. of 7.5 cm². Calculate the current required to produce a flux-density of 1.5 T in the core.

    What would be the value of the flux-density if there was an equivalent of a 1 mm air gap in the ring and current remained the same.

    μ_r = 1000, and μ_o = $4\pi \times 10^{-7}$ H/m.

    (Ans. 1.19 A, 2121000 A/Wb, 0.75 T)

4. (a)   A ring specimen made of steel-alloy has a mean diameter of 20 cm and a cross-sectional area of 6 cm². In a certain magnetic test the reluctance of the ring is found to be 400000 A/Wb. Calculate the relative permeability of the steel-alloy in this case.

    (b)   A radial air gap is now made in the ring and it is uniformly wound with a coil of 250 turns. When a current of 1 A flows in the coil, it is found that the reluctance of the air gap is the same as that for the ring, which is the same as that in (a). Calculate the magnetic flux set up in the ring.

    (Ans. (a) 2080 (b) 313 μ Wb)

5. (a)   Make a neat sketch of a hysteresis loop and with its aid explain the terms:
          (i)   remanent flux density, B_r,
          (ii) Coersive force, H_c,
          (iii) initial magnetisation
          (iv) magnetic saturation.

(b) A ring specimen of cast steel has radial air gap 0.5 mm long cut through the ring and the air gap has a reluctance of 5×10^5 A/Wb. The ring has a mean length of 50 cm and relative permeability of 1000. Calculate the number of turns wound on the ring if 1.5 A is required in these turns to set up a flux of 0.5 mWb.

(Ans. (b) $N = 333$)

6. A ferromagnetic toroid has a mean circumference of 25 cm, a cross sectional area of 2 cm^2 and is wound with a coil of 1000 turns.

A 1 mm radial air gap is cut in the toroid. Calculate the coil current required to establish a flux of 100 μWb in the toroid.

A magnetisation curve for the specimen is given:

B (T)	0.47	0.60	0.66
H (A/m)	2000	3000	4000

(Ans. 0.96 A)

7. (i) Write down an equation for the resistance of a copper wire in terms of its length and cross sectional area, giving the units for each quantity. Give the analogous expression in a magnetic circuit and explain the term of each quantity and the respective units.

(ii) Write down an equation for the magnetomotive force in terms of magnetic flux, magnetic length, permeability and cross sectional area. Give the analogous expression in an electric circuit.

8. A ferromagnetic toroid ring with a mean total length round the iron of 400 mm, the cross-section being 400 mm^2 the number of turns wrapped round uniformly is 100 and the direct current is 5 A. Calculate:

(i) the reluctance of the ring

(ii) the flux in the iron

$$\mu_o \mu_r = 4\pi \times 10^{-4} \text{ H/m}$$

(Ans. (i) 795775 A/Wb, (ii) 628 μWb)

9. In question 8 an air gap of 4 mm is cut, calculate the reluctance in the air gap what is the total reluctance of the circuit? What is the effect of cutting an air gap?

(Ans. 8761560 A/Wb)

5. ELECTROMAGNETISM

5. Applies the fundamental principles and laws governing electromagnetic induction.
a. Explains the motor principles in terms of the interaction between a magnetic field and a current - carrying conductor and applies the relationship F = BIl to simple situations.

THE MAGNETIC EFFECT OF CURRENT

Consider a current-carrying conductor as shown in Fig. 116. A magnetic field is set up around the conductor according to the *corkscrew* rule as shown in the diagram or the right hand gripping rule.

Fig. 116 The magnetic field around a current carrying conductor.

Fig. 117 shows the current flowing out of the paper and Fig. 118 shows the current flowing into the paper.

Fig. 117 The current flowing out of the paper.

Fig. 118 The current flowing into the paper.

The force on the conductor, F, in newtons depends directly on the flux density, B, the length of the conductor, l and the current flowing through the conductor, I,

$$F = BlI.$$

If the current $I = 1$ A, $l = 1$ m, $B = 1$ T then the force on the conductor $F = 1$ newton. If the conductor makes an angle θ with the magnetic field then F is given

$$F = BlI \sin \theta.$$

MOTOR PRINCIPLE

The left hand rule illustrates clearly the motor principle as shown in Fig. 119.

Fig. 119 The Left Hand Rule.

The thumb, the first finger and the middle finger of the left hand when they are stretched out at right angles mutually, we have the force along the thumb, the magnetic field B along the first finger and the current along the middle finger.

If a current - carrying conductor is placed at right angles to a magnetic field as shown in the Fig. 120, the conductor is moved upwards out of the paper with a force $F = BlI$, where B is the flux density set up between the north and south, l is the length of the conductor in the magnetic field, and I is the current flowing through the conductor.

Fig. 120 Magnetic field being by a current - carrying conductor.

The current I generates a force, F, out of the paper. This is the *motor principle*. If the current is reversed then the force acts into paper Fig. 121 and Fig. 122 show the forces into and out of the paper respectively.

Fig. 121 Left Hand Rule.

Fig. 122 Reverse of current.

WORKED EXAMPLE 24

A conductor 45 cm in length and carrying a current of 0.3 A is placed at right angles to a magnetic field having a flux density of 1.0 T. Calculate the force on the conductor.

SOLUTION 24

$$F = BlI$$

where B = flux density in teslas (T) I = current (A)

l = length of conductor (m)

F = force in newtons (N)

$F = 1 \times 0.3 \times 0.45 = 0.135$ N.

*b. Observes the production of induced voltage and describes it as due to change in
  flux linkage.*

Consider the simple situation shown in Fig. 123.

Fig. 123 Induced e.m.f. Lenz's Law.

A stationary coil is connected to a resistor for continuity. A permanent magnet is
moved along the arrow shown towards the coil. The flux linkage, $N\phi$ is changing
whilst the magnet is moving. Due to this change in flux linkage, an induced e.m.f.
e, is established across R as shown. If the magnet moves now in the opposite
direction as shown in Fig. 124, the induced e.m.f. is reversed.

Fig. 124 Induced e.m.f. Lenz's Law.

The induced e.m.f. is directly proportional to the rate of change of flux linkage.

$$e = \frac{d}{dt}(N\phi).$$

Since N is the number of turns of the coil which is constant

$$N\frac{d\phi}{dt}$$

the induced e.m.f. is directly proportional to the rate of change of flux.
If there is no rate of change of flux, there is no induced e.m.f.

c. Establishes the relationship $E = Blv$ and $E = N\dfrac{d\phi}{dt}$ and uses them to solve simple problems.

From the above $E = N\dfrac{d\phi}{dt}$.

Considering a conductor of length l and placed at right angle to a magnetic field as shown in Fig. 125. The conductor cuts the magnetic field with velocity v (m/s).

Fig. 125 A conductor cutting a magnetic field at right angles.

The distance moved in t seconds is given as vt. For a conductor $N = 1$

$$E = N\frac{d\phi}{dt} = N\frac{d(BA)}{dt}$$

$$E = 1\frac{d(BA)}{dt}$$

since $A = l\,(vt) = l\,vt$

$$E = \frac{d}{dt}(B\,l\,v\,t)$$

since B, l and v are constants and t is the only variable, $Blv\dfrac{d}{dt}(t) = Blv$

$$E = Blv$$

E is the induced e.m.f. in the conductor, l, cutting a magnetic field of flux density B with a velocity v.

WORKED EXAMPLE 25

A conductor is moved at 15 m/s through a magnetic field having a flux density of 1 T.

Calculate (a) the e.m.f. induced per metre length of the conductor and

(b) the force per metre length opposing this movement if the conductor is connected in a circuit having a total resistance of 1 Ω.

SOLUTION 25

(a) $E = Blv$

$v = 15$ m/s, $B = 1$ T

the e.m.f. included per metre, $\dfrac{E}{l} = Bv = 1 \times 15 = 15$ V/m.

(b) $F = BIl$

$\dfrac{F}{l}$ = force per metre = $BI = 1 \times 15 = 15$ N/m

where $I = \dfrac{V}{R} = \dfrac{15}{1} = 15$ A.

WORKED EXAMPLE 26

A coil of 1000 turns is subjected to a flux change at the rate of 15 mWb/s. Determine the magnitude of the induced e.m.f.

SOLUTION 26

The magnitude of the induced e.m.f. is given by the formula

$E = N\dfrac{d\phi}{dt}$

$E = 1000 \times 15 \times 10^{-3} = 15$ volts

where $\dfrac{d\phi}{dt} = 15 \times 10^{-3}$ Wb/s.

d. Explains the historical and technical significance of Faraday's and Lenz's Laws.

Fig. 123 establishes an induced e.m.f. such that it opposes the motion. Fig. 124 establishes an induced e.m.f. such that opposes again the motion. Figures 123 and 119 show clearly Lenz's law and this is indicated algebraically by a negative sign

$$E = -N\frac{d\phi}{dt} \quad \dots (1)$$

If L is the self-inductance of the coil in Fig. 123 and Fig. 124, there is an induced e.m.f. in the coil due to the change of current.

$$E = -L\frac{di}{dt} \quad \dots (2)$$

The induced e.m.f. is directly proportional to the rate of change of current. If there were no change in current, the induced e.m.f. would be zero.

Equation (1) and (2) are known as the Faraday's laws of electromagnetism. From equation (2), the self-inductance may be defined

$$\text{Self-inductance} = \frac{\text{induced e.m.f.}}{\text{rate of change of current}}$$

$$L = \frac{E}{\dfrac{di}{dt}}$$

neglecting the negative sign in this case as it only indicates direction of induced e.m.f. or current.

$$E = N\frac{d\phi}{dt} \quad \text{and} \quad E = -L\frac{di}{dt}$$

are the laws due to Faraday and the negative sign is due to Lenz's Law.

WORKED EXAMPLE 27

A centre-tapped coil of 1000 turns is mounted on a magnetic core and has a self-inductance of 50 mH. Calculate:

(i)   the induced e.m.f. when the flux changes at the rate of 40 mWb/s

(ii) the self-inductance of each half of the coil and

(iii) the rate of current change in the coil.

SOLUTION 27

A centre-tapped coil of 1000 turns is shown in Fig. 126.

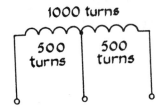

Fig. 126 Centre-tapped coil

(i) $E = N \dfrac{\mathrm{d}\phi}{\mathrm{d}t} = 1000 \times 40 \times 10^{-3} = 40$ volts

(ii) The total self-inductance = 50 mH, the self-inductance of each half coil is 25 mH.

(iii) $E = -L \dfrac{\mathrm{d}i}{\mathrm{d}t}$

re-arranging the formula

$$\frac{\mathrm{d}i}{\mathrm{d}t} = \frac{-40}{50 \times 10^{-3}} = -800 \text{ A/s.}$$

WORKED EXAMPLE 28

A current varying in the manner shown in the diagram of Fig. 127 is passed through a coil having an inductance of 500 mH.

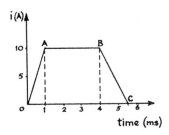

Fig. 127 Current/time graph

Calculate the e.m.f. induced in the coil during the periods *OA*, *AB* and *BC*.

SOLUTION 28

During the period OA

$$E = -L\frac{di}{dt} = -0.5 \times \frac{(10 - 0)}{(1 - 0) \times 10^{-3}} = -5000 \text{ volts}$$

during the period AB

$$E = -0.5\frac{(10 - 10)}{(4 - 1) \times 10^{-3}} = -0.5 \times \frac{0}{3 \times 10^{-3}} = 0 \text{ V}$$

during the period BC

$$E = -0.5\frac{(0 - 10)}{(5.5 - 4) \times 10^{-3}} = \frac{5 \times 10^3}{1.5} = 3333 \text{ volts.}$$

This example clearly established Lenz' Law.

e. *Describes the concept of eddy currents and eddy current loss, explaining their significance under conditions of a.c. magnetisation.*

Coil normally are wound on ferromagnetic materials in order to increase the self-inductance. If the ferro-magnetic materials are solid, eddy currents flow in the solid material which cause a power loss and thus decreases the efficiency of the device.

To reduce these "Eddy currents", the resistance of the core is increased by replacing it with thin painted laminations.

Eddy currents are objectionable in generator and motor armature and transformer cores, in which they cause power losses and hence the efficiencies are decreased.

Understands the concepts of self and mutual inductance and relates these to the transformer principle.

f. *Defines self-inductance of a coil in terms of the proportionality of flux linkages and current in a linear magnetic medium and describes the production of induced voltage due to change in flux linkages.*

SELF INDUCTANCE OF COIL

$$e = -L\frac{di}{dt} = -N\frac{d\phi}{dt}$$

$$L = \frac{e}{\dfrac{di}{dt}} = \frac{\text{induced e.m.f.}}{\text{rate of change of current}}$$

L = self-inductance (henrys)

at steady state

$$LI = N\phi$$

$$L = \frac{N\phi}{I}.$$

The self-inductance is directly proportional to the flux linkages $(N\phi)$ and inversely proportional to the current.

WORKED EXAMPLE 29

The flux linked with a coil changes steadily from 5 mWb to 75 mWb in 7 ms. The average value of induced e.m.f. is 150 V. How many turns are there on the coil?

SOLUTION 29

$e = N\dfrac{d\phi}{dt}$ = the average induced e.m.f. (neglecting the negative sign)

$$150 = N\frac{(75 \times 10^{-3} - 5 \times 10^{-3})}{7 \times 10^{-3}} = N\frac{70}{7} = 10\,N$$

$N = 15$ the number of turns.

WORKED EXAMPLE 30

A current of 200 mA flowing in a 75 mH coil is reversed linearly in 150 ms. Calculate the value of the e.m.f. induced in the coil.

SOLUTION 30

$$E = -L\frac{di}{dt} = -75 \times 10^{-3}\frac{(-200-200)}{150 \times 10^{-3}} \times 10^{-3}$$

$$E = \frac{75 \times 400}{150} \times 10^{-3} = 0.2 \text{ volts.}$$

b. Deduces and applies the relationship $L = \dfrac{N^2}{S}$ and $E = L\dfrac{di}{dt}$

DESIGNING A COIL

$$L = \frac{N\phi}{I} \text{ at steady state, } H = \frac{IN}{l} \text{ and } S = \frac{l}{\mu_o \mu_r A}$$

$$L = \frac{NBA}{I} = \frac{NBAN}{Hl} = \frac{N^2 BA}{Hl} = N^2 \mu_o \mu_r \frac{A}{l} = \frac{N^2}{\dfrac{l}{\mu_o \mu_r A}} = \frac{N^2}{S}$$

the self-inductance depends greatly on the number of turns

$$L = \frac{N^2}{S}$$

and it is inversely proportional to the reluctance.

$$L = N^2 \mu_o \mu_r \frac{A}{l}$$

Since μ_o, A and l are constants then the self-inductance depends on N^2 and the type of material with relative permeability, μ_r. A coil wound on a ferro-magnetic material gives a much higher self-inductance than when it is wound on a non-magnetic material.

c. Defines mutual inductance and describes the production of induced voltage due to change of mutual flux linkage.

MUTUAL INDUCTANCE BETWEEN TWO COILS

Mutual inductance only exists between two coils. Consider two coils A and B which are mutually coupled as shown in Fig. 128, where M is the mutual inductance.

Fig. 128 Two coils closely coupled with mutual inductance M.

Two coils are said to have mutual inductance of 1 henry if an e.m.f. of 1 volt is induced in the secondary when the primary current changes at the rate of 1 ampere per second.

When the switch S is closed, the current in coil A changes, the induced e.m.f. in coil B is given as

$$E_2 = M\frac{di_1}{dt} = N_2\frac{d\phi_2}{dt}.$$

WORKED EXAMPLE 31

Two coils A and B are mutually coupled, the mutual inductance between them being 0.25 H. If the coils have 500 turns each, determine the induced e.m.f. in coil B when   (i)   the current in coil A changes at 75 A/s
(ii)   the flux linking the coil changes at 0.5 Wb per second.

SOLUTION 31

(i) $E_2 = M\dfrac{di_1}{dt} = 0.25 \times 75 = 18.75$ volts

(ii) $E_2 = N\dfrac{d\phi_2}{dt} = 500 \times 0.5 = 250$ volts.

COILS CONNECTED IN SERIES AIDING

Two coils A and B are connected on an iron former as shown in Fig. 129 in series aiding

Fig. 129   Two coils connected in series aiding.

If the self-inductance of A is L_A and the self-inductance of B is L_B, the total self-inductance is given by

$$L = L_A + L_B + 2M$$

where M is the mutual inductance of the two coils.

COILS CONNECTED IN SERIES OPPOSING

Two coils A and B are connected on an iron former as shown in Fig. 130.

Fig. 130 Two coils connected in series opposing.

If again the self-inductance of A is L_A and the self-inductance of B is L_B, the total self-inductance is given by
$$L = L_A + L_B - 2M.$$

WORKED EXAMPLE 32

Two coils, A and B, have self-inductances of 10 mH and 20 mH respectively. When connected in series aiding the total inductance is 35 mH. Calculate the mutual inductance and estimate the total inductance of the coils when they are connected in series opposing.

SOLUTION 32

Referring to Fig. 129

$L_A = 10$ mH and $L_B = 20$ mH

$L = L_A + L_B + 2M$

$35 \times 10^{-3} = 10 \times 10^{-3} + 20 \times 10^{-3} + 2M$

$(35 - 30) \times 10^{-3} = 2M$

$M = 2.5$ mH

Referring to Fig. 130

$L = L_A + L_B - 2M = 10 \times 10^{-3} + 20 \times 10^{-3} - 2 \times 2.5 \times 10^{-3}$

$\quad = 30 \times 10^{-3} - 5 \times 10^{-3}$

$L = 25 \times 10^{-3}$ H

therefore

$L = 25$ mH

INDUCTIVE AND NON INDUCTIVE CIRCUITS

Straight copper wires possess a very small self-inductance, if the copper wires are in the form of a coil, the self-inductance is more predominant.

To increase the self-inductance the coil is wound on a magnetic circuit.

Therefore straight copper wires, coils on non-magnetic and coils on magnetic circuits are inductive.

Non-inductive circuits such as high stability resistors and incandescent 1 A are made of wires bent back on itself since the field due to current flowing in one direction cancels that due to current flowing in opposite direction Fig. 131 shows the construction of a non-inductive resistor.

Fig. 131   Non-inductive resistor.   Wire is bent back on itself.

d. Describes the transformer principle in terms of Lenz's Law and induced volts per turn, deducing the effect of turns ratio on voltage ratio.

THE TRANSFORMER PRINCIPLE

Two coils of turns N_p and N_s are wound on a ferro-magnetic core as shown in Fig. 132.

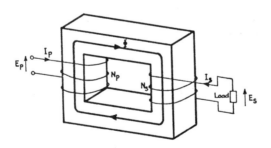

Fig. 132   The basic principle of a transformer.

The coil N_p is connected to an a.c. supply E_p and the coil N_s is connected to a load. The former coil is called *the primary* and the latter coil is called *the secondary*.

95

Applying Faraday's laws of electromagnetism

$$E_p = -N_p \frac{\mathrm{d}\phi}{\mathrm{d}t} \quad \dots (1) \quad \text{and} \quad E_s = -N_s \frac{\mathrm{d}\phi}{\mathrm{d}t} \quad \dots (2)$$

assuming that the rate of change of flux is the same, that is, no flux leakage occurs. Dividing equation (1) by equation (2)

$$\frac{E_p}{E_s} = \frac{-N_p \frac{\mathrm{d}\phi}{\mathrm{d}t}}{-N_s \frac{\mathrm{d}\phi}{\mathrm{d}t}} = \frac{N_p}{N_s}$$

$$\frac{E_p}{E_s} = \frac{N_p}{N_s} \quad \dots (3)$$

Therefore, the ratio of the induced e.m.f. of the primary to the induced e.m.f. of the secondary is equal to the turns ratio.

Assuming no power loss between the primary and secondary, the primary volt-amperes are equal to the secondary volt-amperes

$$E_p I_p = E_s I_s$$

or
$$\frac{E_p}{E_s} = \frac{I_s}{I_p} \quad \dots (4)$$

combining equation (3) and (4), we have

$$\frac{E_p}{E_s} = \frac{N_p}{N_s} = \frac{I_s}{I_p} \quad \dots (5)$$

---

e. *Deduces that energy stored in an inductor is* $\frac{1}{2} L I^2$.

Power is the rate of doing work

P = power in watts

W = energy or work in joules

$$P = \frac{W}{t} \quad \text{or} \quad W = Pt = EIt = IRIt = I^2 Rt.$$

When an inductor is completely demagnetised and is connected across a d.c. supply, the current initially is zero and hence the energy is zero, t seconds later the inductor is energised and the current has grown to a

finite value i, the induced e.m.f. is $L\dfrac{di}{dt}$ and its energy will be $\left(L\dfrac{di}{dt}\right)i\,dt$.

The energy stored in the inductor $= \displaystyle\int_0^I \left(L\dfrac{di}{dt}\right)i\,dt = \int_0^I Li\,di$

$$= \left[\dfrac{1}{2}Li^2\right]_0^I = \dfrac{1}{2}LI^2.$$

The current grows from $i = 0$, to $i = I$, the final steady current.

The energy stored in an inductor from $i = 0$ to $i = I$ is $W = \dfrac{1}{2}LI^2$, the energy stored in an inductor at any instant is $W = \dfrac{1}{2}Li^2$ where i is the instantaneous current.

WORKED EXAMPLE 33

(i) Calculate the e.m.f. induced in an inductance in an inductance of 500 mH when the current is changing at a rate of 200 A/s.

(ii) Calculate the energy stored in this inductance when the current is 5 A.

SOLUTION 33

(i) $E = -L\dfrac{di}{dt} = -500 \times 10^{-3} \times 200 = -100$ volts

(ii) $W = \dfrac{1}{2}Li^2 = \dfrac{1}{2} \times 500 \times 10^{-3} \times 5^2 = 6.25$ joules.

WORKED EXAMPLE 34

Calculate the energy contained in an inductance of 5 H carrying a current of 1 A.

SOLUTION 34

$W = \dfrac{1}{2}LI^2 = \dfrac{1}{2} \times 5 \times 1^2 = 2.5$ J.

f. Solves problems on self-inductance, mutual inductance and the transformer principle.

THE RESISTANCE VIEWED AT THE PRIMARY OF A TRANSFORMER

Consider a simple transformer in Fig. 133 whose a.c. input resistance is required

Fig. 133 The a.c. input resistance at the primary of the transformer.

From the equation $\dfrac{E_1}{E_2} = \dfrac{N_1}{N_2} = \dfrac{I_2}{I_1}$ we have

$$E_1 = \left[\frac{N_1}{N_2}\right] E_2 \ \dots \ (1) \text{ and } I_1 = I_2 \left[\frac{N_2}{N_1}\right] \ \dots \ (2)$$

and load $R = \dfrac{E_2}{I_2} \ \dots \ (3)$

The a.c. input resistance viewed at the primary terminals, R_{in} is given

$$R_{in} = \frac{E_1}{I_1} = \frac{\text{Input voltage}}{\text{Input current}}$$

substituting in this equation, equation (1) and (2)

$$R_{in} = \frac{E_2 \left[\dfrac{N_1}{N_2}\right]}{I_2 \left[\dfrac{N_2}{N_1}\right]}$$

$$R_{in} = \frac{E_2}{I_2} \left[\frac{N_1}{N_2}\right]^2$$

98

and finally substituting $R = \dfrac{E_2}{I_2}$, we have

$$R_{in} = \left(\dfrac{N_1}{N_2}\right)^2 R \ \dots \ (4)$$

the resistance viewed at the primary.

WORKED EXAMPLE 35

An ideal transformer having a 50:1 step down ratio, has a load current and voltage of 0.2 A and 50 V respectively. Calculate the primary current and voltage.

SOLUTION 35

Using the equation

$$\dfrac{I_1}{I_2} = \dfrac{N_2}{N_1}$$

or $\quad I_1 = I_2 \dfrac{N_2}{N_1}$

$$I_1 = 0.2 \times \dfrac{1}{50} = 4 \text{ mA}.$$

Using equation

$$\dfrac{E_1}{E_2} = \dfrac{N_1}{N_2}$$

or $\quad E_1 = \dfrac{N_1}{N_2} E_2 = 50 \times 50 = 2500 \text{ V}.$

WORKED EXAMPLE 36

A 5:1 step down transformer has a primary voltage supply of 240 V and a secondary load resistance of 1000 Ω. Calculate:

(i)   the resistance at the primary terminals

(ii)   the secondary voltage and

(iii)   the primary and secondary currents.

SOLUTION 36

(i) $\quad R_{in} = \left[\dfrac{N_1}{N_2}\right]^2 R$

$\quad R_{in} = 5^2 \times 1000 = 25000\ \Omega$

(ii) $\quad E_2 = E_1 \dfrac{N_2}{N_1} = 240 \times \dfrac{1}{5} = 48$ volts

(iii) $\quad I_2 = \dfrac{E_2}{R} = \dfrac{48}{1000} = 48$ mA

$\quad I_1 = \dfrac{E_1}{R_{in}} = \dfrac{240}{25000} = 9.6$ mA.

WORKED EXAMPLE 37

A coil is wound with 1000 turns. A current of 2 amperes flowing in the coil produces a flux of 100 μWb. Determine the value of the self-inductance.

SOLUTION 37

At steady state

$\quad LI = N\phi \quad \dots (1)$

the self-inductance $L = \dfrac{N\phi}{I}$

$$L = \frac{1000 \times 100 \times 10^{-6}}{2}$$

$\quad L = 50$ mH.

WORKED EXAMPLE 38

A coil having an inductance of 225 mH is carrying a current of 20 A. Calculate the induced e.m.f. in the coil when the current is:-
(i)   reduced to zero in 0.05 seconds
(ii)   reversed in 0.05 seconds
(iii)   increased to 50 A in 3 seconds.

SOLUTION 38

(i) $E = -L\dfrac{di}{dt} = -225 \times 10^{-3} \times \dfrac{(0-20)}{0.05} = 90$ V

(ii) $E = -L\dfrac{di}{dt} = -225 \times 10^{-3} \times \dfrac{(-20-20)}{0.05} = 180$ V

(iii) $E = -L\dfrac{di}{dt} = -225 \times 10^{-3} \times \dfrac{(50-20)}{3} = 2.25$ V.

SUMMARY 5

Left and Right Hand Rules are shown in Fig. 134 and Fig. 135.

Fig. 134 Left Hand Rule   Fig. 135 Right Hand Rule

1. $F = BIl \sin \theta$ F = force on a conductor (N)

2. $E = Blv$ B = flux density (T)

3. $E = -L\dfrac{di}{dt}$ I = Current (A)

4. $E = -N\dfrac{d\phi}{dt}$ l = length of conductor (m)

101

5. $LI = N\phi$ steady state

θ = angle between the straight conductor and magnetic field (degrees)

6. $E_2 = M\dfrac{di_1}{dt}$

E = induced e.m.f. (V)

7. $E_2 = N_2\dfrac{d\phi_2}{dt}$

v = velocity (m/s)

8. $W = \dfrac{1}{2}LI^2$

L = self-inductance (H)

9. $\dfrac{E_p}{E_s} = \dfrac{N_p}{N_s}$

$\dfrac{di}{dt}$ = rate of change of current (A/s)

10. $\dfrac{E_p}{E_s} = \dfrac{I_s}{I_p}$

$\dfrac{d\phi}{dt}$ = rate of change of flux (Wb/s)

11. $R_{in} = \left(\dfrac{N_1}{N_2}\right)^2 R_{load}$

N = number of turns

M = mutual inductance (H)

W = energy stored (J)

E_p = primary voltage (V)

E_s = secondary voltage (V)

R_{in} = a.c. resistance at the input terminals of the transformer (Ω)

$\dfrac{N_1}{N_2}$ = turns-ratio

R_{load} = load resistance (Ω).

EXERCISES 5

1. The force on a conductor of 0.5 m is 0.25 N and the conductor carries a current of 50 mA and is placed at right angles to a magnetic field. Determine the flux density of the magnetic field. (Ans. 10 T).

2. A conductor 40 cm in length and carrying a current of 150 mA is placed at 80° to a magnetic field having a flux density of 0.9 T. What is the angle of the conductor to the magnetic field having the same flux density and the force on the conductor is 1/5 that of the first case. (Ans. 11.32°).

3. A conductor at right angles to a uniform magnetic field of flux density, 50 mT has 10 cm of its length in the field. Calculate the current required to cause a force of 100 mN to act on the conductor. (Ans. 20 A).

4. A rectangular coil 20 cm long and 15 cm wide and consisting of one turn is rotated about its longer axis at 500 r.p.m. in a uniform magnetic field of 1 T field is situated normal to the axis. Determine the e.m.f. induced in the single turn coil, and in a similar coil of 100 turns. (Ans. 0.79 V, 79 V).

5. An e.m.f. of 5 volts is induced in a coil due to a flux change at the rate of 0.5 Wb/s. Calculate the number or turns of the coil. (Ans. 10).

6. Calculate the e.m.f. induced in an inductance of 200 mH when the current is changing at a rate of 500 A/s. (Ans. 100 V).

7. With the aid of two coils closely coupled together explain what is meant by mutual inductance. Two coils A and B are mutually coupled, the mutual inductance between them being 0.2 H. Calculate the e.m.f. in coil B when the current in coil A changes at 100 A/s. (Ans. 20 V).

8. An ideal step up transformer having a turns ratio of 1:10 is connected to a supply voltage of 240 V. The secondary load is a resistor of 12 KΩ. Calculate: (i) the secondary voltage and current
    (ii) the primary current
    (iii) the resistance viewed at the primary terminals.
    (Ans. (i) 2400 V, 0.2 A (ii) 2 A, (iii) 120 Ω).

9. (i) Show that the self-inductance of a coil is directly proportional to the number of turns squared and the permeability of the magnetic material.
    (ii) Show that the reluctance of non-magnetic material is constant for all values of current.
    (iii) Show that the reluctance of a magnetic material varies with the current.

10. Explain briefly and clearly the construction of inductive and non-inductive components using conducting wires.

6. ALTERNATING VOLTAGES AND CURRENTS

6. Displays waveforms and determines the main parameters used to describe and measure them.

a. Defines the terms amplitude, period, frequency, instantaneous, peak-to-peak r.m.s. average in relation to alternating (sinusoidal and non-sinusoidal) and unidirectional waveforms and uses an oscilloscope to display and measure these parameters.

ALTERNATING SINUSOIDAL WAVEFORMS

Fig. 136 shows an alternating sinusoidal waveform of two cycles, the *y*-axis is *A* sin *ωt* and the *x*-axis is *ωt*.

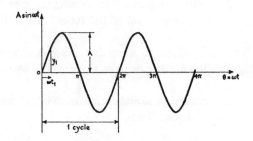

Fig. 136 A.C. sinusoidal waveform.

Trigonometrically this waveform is represented by the equation

$$y = A \sin \omega t.$$

The waveform commences at the origin O.

AMPLITUDE

The maximum or peak value is called the amplitude and it is represented by A.

PERIOD

The period is the time taken to trace out one complete cycle and it is represented by T.

FREQUENCY

The frequency is the number of cycles traced out in one second and is represented by f.

INSTANTANEOUS

The instantaneous value of y is the value at a certain instant t, (t_1 for example). At t_1, $y = y_1 = A \sin \omega t_1$, y_1 is the instantaneous value of y at $t = t_1$.

PEAK TO PEAK

The peak-to-peak value is the value between the maximum or peak positive value and the maximum of peak negative value. In this case, the peak-to-peak value 2 A.

R.M.S. (ROOT MEAN SQUARE)

The r.m.s is the square root value of the mean squares. This is best illustrated as follow:

$\theta°$	0	30	60	90	120	150	180
$A \sin \theta°$	0	$\frac{1}{2}A$	$\frac{\sqrt{3}}{2}A$	A	$\frac{\sqrt{3}}{2}A$	$\frac{1}{2}A$	0
$A^2 \sin^2 \theta°$	0	$\frac{1}{4}A^2$	$\frac{3}{4}A^2$	A^2	$\frac{3}{4}A^2$	$\frac{1}{4}A^2$	0

$\theta°$	210	240	270	300	330	360
$A \sin \theta°$	$-\frac{1}{2}A$	$-\frac{\sqrt{3}}{2}A$	$-A$	$-\frac{\sqrt{3}}{2}A$	$-\frac{1}{2}A$	0
$A^2 \sin^2 \theta°$	$\frac{1}{4}A^2$	$\frac{3}{4}A^2$	A^2	$\frac{3}{4}A^2$	$\frac{1}{4}A^2$	0

The root mean square

$$= \sqrt{\frac{0^2 + \frac{1}{4}A^2 + \frac{3}{4}A^2 + A^2 + \frac{3}{4}A^2 + \frac{1}{4}A^2 + 0 + \frac{1}{4}A^2 + \frac{3}{4}A^2 + A^2 + \frac{3}{4}A^2 + \frac{1}{4}A^2}{12}}$$

$$= \sqrt{\frac{6A^2}{12}} = \frac{A}{\sqrt{2}} = 0.707 \text{ A approximately.}$$

AVERAGE

The average value of one complete cycle is zero since the positive area of the half cycle cancels the negative area of the other half cycle.

The average value of one half cycle is given as $\dfrac{2A}{\pi}$ = 0.637 A approximately.

NON SINUSOIDAL WAVEFORMS

Non sinusoidal waveforms such as the triangular and square waveforms are shown in Fig. 137 and Fig. 138 respectively.

Fig. 137 Triangular waveforms.

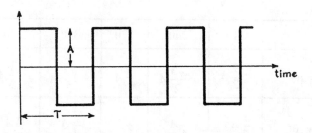

Fig. 138 Square waveforms.

These wave forms are called complex or non-sinusoidal waves. The amplitude and the period are shown on the diagrams.

b. Defines form factor and determines the approximate average and r.m.s. value of given sinusoidal and non-sinusoidal waveforms.

A sinusoidal waveform

Form factor $= \dfrac{\text{r.m.s. value}}{\text{average value}} = \dfrac{0.707A}{0.637\ A} = 1.11.$

106

For any sinusoidal current or voltage the form factor is the same and it is always 1.11 approximately to three significant figures.

For a non-sinusoidal waveform the form factor is different and should be determined for each waveform.

Use phasor and algebraic representation of sinusoidal quantities.
a. Defines a phasor quantity.

PHASOR OR PHASOR DIAGRAM

A phasor quantity has magnitude and direction.

A current in a circuit is 3 A and leads the voltage by 25°. If the voltage is taken as a reference, a horizontal line of indefinite magnitude, then three units are scaled along a positive direction of 25° to the voltage. If the current lags by 25°, then three units are scaled along a negative direction of 25°. Fig. 139 represents the phasor quantity of current with respect to the voltage.

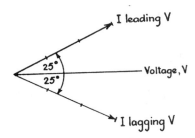

Fig. 139 Leading and Lagging currents.

b. Determines the resultant or the addition of two sinusoidal voltages by graphical and phasor representation.

PHASOR REPRESENTATION

WORKED EXAMPLE 39

Two sinusoidal voltages are applied across two components in a circuit

$$E_1 = 10 \sin \omega t$$

$$E_2 = 20 \sin \left(\omega t + \pi/2 \right).$$

Determine the resultant total voltage across the circuit and express in the form
$E_R = E_m \sin \left(\omega t \pm \phi \right).$

107

SOLUTION 39

Since E_1 is a reference phasor quantity, it is drawn 10 units horizontally and since E_2 is 90° leading or $\dfrac{\pi}{2}$ leading, it is drawn up perpendicularly 20 units as shown in Fig. 140.

Fig. 140 Phasor representation of voltages E_1 and E_2 and their resultant E_R.

The resultant in this case is the hypotenuse of the right angle triangle.

$$E_m^2 = E_{1m}^2 + E_{2m}^2$$

$$E_m = \sqrt{20^2 + 10^2} = 22.36.$$

From the right-angle triangle OE_1E_r, $\tan \theta = \dfrac{20}{10} = 2$ and hence $\theta = 63.43°$.

The resultant is given by

$$E_R = 22.36 \sin\left(\omega t + 63.43°\right)$$

but ω is the angular velocity in radians per second, ωt is the angle in radian and it is better to express θ in radians.

$$63.43° = 63.43° \times \frac{\pi}{180°} = 1.11^c$$

$$E_R = 22.4 \sin\left(\omega t + 1.11^c\right)$$

the resultant voltage in three significant figures.

GRAPHICAL REPRESENTATIONS

WORKED EXAMPLE 40

If $E_1 = 10 \sin \omega t$ and $E_2 = 20 \sin\left(\omega t + \pi/2\right)$ determine graphically $E_1 + E_2$.

SOLUTION 40

θ^c	0	$\frac{\pi}{6}$	$\frac{\pi}{3}$	$\frac{\pi}{2}$	$2\frac{\pi}{3}$	$5\frac{\pi}{6}$	π	$7\frac{\pi}{6}$	$4\frac{\pi}{3}$	$3\frac{\pi}{2}$	$5\frac{\pi}{3}$	$11\frac{\pi}{6}$	2π
$\sin\theta^c$	0	0.5	0.866	1	0.866	0.5	0	-0.5	-0.866	-1	-0.866	-0.5	0
$E_1 = 10\sin\theta^c$	0	5	8.66	10	8.66	5	0	-5	-8.66	-10	-8.66	-5	0
$\sin\left(\theta^c + \frac{\pi}{2}\right)$	1	0.866	0.5	0	-0.5	-0.866	-1	-0.866	-0.5	0	0.5	0.866	1
$E_2 = 20\sin\left(\theta^c + \frac{\pi}{2}\right)$	20	17.32	10	0	-10	-17.32	-20	-17.32	-10	0	+10	17.32	20
$E_1 + E_2$	20	22.3	18.7	10	-1.34	-12.3	-20	-22.3	-18.7	-10	1.34	12.3	20

Fig. 141 shows the plot of the resultant waveform, $E_1 + E_2$ against ωt.

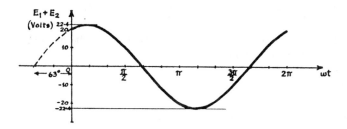

Fig. 141 Resultant waveform for E_1 and E_2, leading by $63°$

The amplitude of the resultant waveform is approximately 22.3 and the waveform lead by about 1.1^c.

c. Explains the phase-angle relationship between two alternating quantities.

PHASE ANGLE LEADING AND LAGGING

Fig. 142 shows a lagging, C, and a leading, B, waveforms with respect to the reference, A, waveform.

109

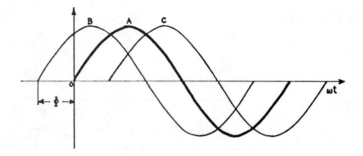

Fig. 142 Reference, leading and lagging waveforms.

A is taken as a reference waveform since it commences at the origin.

B is leading *A* since peak value of the waveform is reached before that *A*, *C* is lagging *A* since the peak value of the waveform is reached after that of *A*.

d. Defines a sinusoidal voltage in the form $v = V_m \sin(\omega t + \phi)$.

The instantaneous value of the voltage is given in terms of the maximum value, the angular velocity, time and phase angle.

$v = V_m \sin(\omega t \pm \phi)$ volts

v = instantaneous value of the voltage at time t

V_m = the peak or maximum value of voltage

ω = the angular velocity in radians per second

t = time in seconds

ϕ = phase angle in radians.

When the sign in front of ϕ is positive then the waveform leads the reference, when the sign in front of ϕ is negative then the waveform lags the reference.

e. Determines current from the application of sinusoidal voltage to a resistive circuit.

If a sinusoidal voltage is applied across a resistive circuit, the current through the circuit is sinusoidal.

If $v = V_m \sin(\omega t \pm \phi)$ then $i = \dfrac{V}{R}$ or

$$i = \frac{V_m}{R} \sin(\omega t \pm \phi) = I_m \sin(\omega t \pm \phi).$$

110

f. Interrelates graphical, phasor and algebraic representation in the determination of amplitude, instantaneous value, frequency period and phase of sinusoidal voltage and currents.

WORKED EXAMPLE 41

An alternative current is given by the expression $i = 10 \sin \left(3142t + \dfrac{\pi}{4} \right)$ A.

Calculate (i)   the peak value

             (ii) the average value

             (iii) the r.m.s. value

             (iv) the frequency

             (v) the periodic time

             (vi) the angular velocity

             (vii) the instantaneous value of the current when $t = 0.25$ ms.

SOLUTION 41

(i) $I_{max} = 10$ A

(ii) $I_{av} = 10 \times \dfrac{2}{\pi} = 6.37$ A

(iii) $I = \dfrac{10}{\sqrt{2}} = 7.07$ A

(iv) $f = \dfrac{\omega}{2\pi} = \dfrac{3142}{2\pi} = 500$ Hz

(v) $T = \dfrac{1}{f} = \dfrac{1}{500} = 2$ ms

(vi) $\omega = 3142$ radians per second

(vii) $i = 10 \sin \left(3142 \times 0.25 \times 10^{-3} + \dfrac{\pi}{4} \right)$

      $= 10 \sin \left(0.7854^c + 0.7854^c \right)$

      $= 10 \sin \left(1.5708^c \right)$

      $= 10$ A.

WORKED EXAMPLE 42

An instantaneous voltage is given by the expression $v = 100 \sin \left(3142t + \dfrac{\pi}{6} \right)$.

(a) Calculate: (i)   the frequency
          (ii) the period
          (iii) the r.m.s. value
          (iv) the average value
          (v)   the form factor
          (vi) the instantaneous value after 1 ms.

(b) Represent the above voltage and the instantaneous current $i = 100 \sin 3142t$ A in phasor diagram.

SOLUTION 42

(a) (i) $f = \dfrac{\omega}{2\pi} = \dfrac{3142}{2\pi} = 500$ Hz

(ii) $T = \dfrac{1}{f} = \dfrac{1}{500} = 2$ ms

(iii) $V_{r.m.s.} = \dfrac{V_{max}}{\sqrt{2}} = \dfrac{100}{\sqrt{2}} = 70.7$ volts

(iv) $V_{av} = \dfrac{2}{\pi} V_{max} = 2 \times \dfrac{100}{\pi} = 63.7$ volts

(v) Form factor $= \dfrac{\text{r.m.s.}}{\text{average}} = \dfrac{70.7}{63.7} = 1.11$

(vi) $v = 100 \sin \left(3142 \times 1 \times 10^{-3} + \dfrac{\pi}{6} \right) = 100 \sin \dfrac{7\pi}{6}$

$= 100 \left(-\dfrac{1}{2} \right) = -50$ volts.

(b) (i) $i = 100 \sin 3142\,t$ A

this is taken as a reference since we draw a horizontal phasor of 100 V equivalent to 5 cm (that is 20 V per cm is the scale). The voltage is drawn at 30° to the horizonal and equal to 5 cm.

Fig. 143 represents the current and voltage in a phasor diagram.

Fig. 143 v leads i by 30° or $\dfrac{\pi}{6}$.

WORKED EXAMPLE 43

(a) Write down the mathematical expression representing a sinusoidal current having an r.m.s. value of 250 mA and frequency 2500 Hz and leading a voltage by a phase angle of 75°.

(b) Write down the mathematical expression representing a sinusoidal voltage having a peak value of 100 V and a periodic time of 2.5 ms and lagging a current by a phase angle of $\dfrac{\pi}{9}$.

SOLUTION 43

(a) $i = 250 \times 10^{-3}\sqrt{2} \sin \left(2\pi\,2500t + \dfrac{5\pi}{12} \right)$ A.

where $75° = 75° \times \dfrac{\pi}{180}^° = \dfrac{5\pi^c}{12}$

$i = 0.3535 \sin \left(15708t + \dfrac{5\pi}{12} \right)$ A

(b) $v = 100 \sin \left(2\pi\,\dfrac{1000}{2.5}t - \dfrac{\pi}{9} \right)$ V

where $T = \dfrac{1}{f} = 2.5$ ms

$f = \dfrac{1}{2.5 \times 10^{-3}} = \dfrac{1000}{2.5} = 400$ Hz

$v = 100 \sin \left(2513.3t - \dfrac{\pi}{9} \right)$.

113

g. Determines power in an a.c. resistive circuit from given data.

Power in a resistive a.c. circuit is given by the r.m.s. value of current squared times the resistance of the resistor

$$P = I^2 R$$

where I = r.m.s. value of current

or $P = \dfrac{V^2}{R}$

where V = r.m.s. value of voltage.

The power absorbed by a resistive load in an a.c. circuit is given simply by the product of the r.m.s. values of voltage and current.

$$P = IV$$

where I and V are the r.m.s. values of current and voltage.

Since the phase difference between the voltage and current waveforms in an a.c. resistive circuit is zero then cos 0 = 1 and $P = IV$. But the phase difference between I and V in non-resistive components is finite and this affects the power absorbed by the load. This will be covered in detail in the next chapter.

If $v = V_m \sin \omega t$ is applied to a circuit of resistance R Ω, the resulting current is represented by $i = I_m \sin \omega t$.

The power at any instant $= P = IV = I_m \sin \omega t \, V_m \sin \omega t$.

Mean power over a cycle $=$ (mean value of i^2) $\times R$

$=$ (mean value of $I_m^2 \sin^2 \omega t$) $\times R$

$= \dfrac{I_m^2}{2} R = \dfrac{I_m}{\sqrt{2}} \times \dfrac{I_m}{\sqrt{2}} \times R$

$P = I \times I \times R = I^2 R = VI.$

Fig. 144 shows the instantaneous values of current, voltage and power

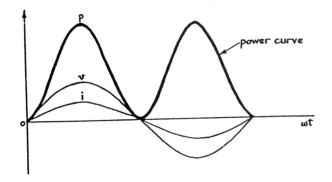

Fig. 144 Power curve for an a.c. resistive curve.

Equation of power curve is given by

$$p = iv = I_m V_m \sin^2 \omega t = \frac{I_m V_m}{2}(1 - \cos 2\omega t).$$

WORKED EXAMPLE 44

A current wave of equation $i = 10 \sin \omega t$ mA is flowing through a 1000 Ω non-inductive resistor. Calculate the p.d. across the circuit and the instantaneous value of the power at $\omega t = \frac{\pi}{2}$.

SOLUTION 44

The p.d. across the 1 KΩ resistor $v = i \times 1000 \sin \omega t$ mV

$$= 10 \sin \omega t \text{ V}$$

The instantaneous value of the power $= p = iv$

$$p = 10 \times 10^{-3} \sin \omega t \times 10 \sin \omega t$$

$$= 0.1 \sin^2 \omega t$$

when $\omega t = \frac{\pi}{2}$, $\sin^2 \omega t = \sin^2 \frac{\pi}{2} = 1$ and $p = 0.1$ W.

SUMMARY 6

$$i = I_m \sin \left(\omega t \pm \phi \right)$$

$$v = V_m \sin \left(\omega t \pm \phi \right)$$

i, v are instantaneous values of current and voltage respectively

I_m, V_m are the peak values of current and voltage respectively

I, V are the r.m.s. values of current and voltage respectively

$$I = \frac{I_m}{\sqrt{2}} \text{ and } V = \frac{V_m}{\sqrt{2}}$$

$\omega = 2\pi f$ = angular velocity radians per second

f = frequency in Hz

$T = \dfrac{1}{f}$ periodic time in seconds

ϕ = phase angle in radians.

Positive phase angle (leading)
Negative Phase angle (lagging)

Form factor = $\dfrac{\text{r.m.s. value}}{\text{average value}}$

Form factor = 1.11 for sinusoidal waves.

Average value of current or voltage

$$I_{av} = \frac{2}{\pi} I_m \qquad\qquad V_{av} = \frac{2}{\pi} V_m.$$

$$P = IV = I^2 R = \frac{V^2}{R} \text{ Power in a resistor.}$$

EXERCISES 6

1. (i) Sketch one cycle of a non-sinusoidal waveform having a peak value of 50 volts and a period of 5 ms.

   (ii) Sketch one cycle of a sinusoidal current waveform having a 100 mA r.m.s. value and frequency of 1000 Hz.

   (iii) Sketch one cycle of a sinusoidal voltage waveform having a peak value of 100 volts, a frequency of 50 Hz and a phase lag of $30°$.

2. Draw phasors to represent the following voltages $v_1 = 15 \sin \omega t$,

   $v_2 = 25 \sin \left(\omega t - \dfrac{\pi}{4} \right)$. Determine by calculation the resultant voltage

   $v_1 + v_2$, in the form $V_R = V_{max} \sin (\omega t \pm \phi)$.

   (Ans. 37.15, 0.496^c, $v_R = 37.15 \sin (\omega t - 0.496^c)$).

3. A sinusoidal alternating voltage has the trigonometric equation

   $v = 100 \sqrt{2} \sin \left(377 t + \dfrac{\pi}{4} \right)$ V

   Determine: (i) the peak value
   (ii)   the average value
   (iii) the r.m.s. value
   (iv)   the angular velocity
   (v)   the frequency
   (vi)   the periodic time
   (vii) the instantaneous voltage when $t = 5$ ms
   (viii) the phase angle, state if it lags or leads.

   (Ans. (i) 141.4 V   (ii) 90 V   (iii) 100 V   (iv) 377 rad/s

   (v) 60 Hz   (vi) 16.7 ms   (vii) 64.2 V   (viii) $\dfrac{\pi}{4}$ leading).

4. An alternating voltage, $v = 10 \sin \left(500 \pi t - \dfrac{\pi}{6} \right)$ is applied across a 100 Ω resistor, write down an expression for the instantaneous current. Sketch one cycle of the current and voltage on the same time base.

   (Ans. $i = 100 \sin \left(500 \pi t - \dfrac{\pi}{6} \right)$ mA)

117

5. An instantaneous current in a circuit is given by the expression

$$i = 100 \sin \left(314200t - \frac{\pi}{12} \right) \text{ mA.}$$

(a) Calculate: (i) its frequency
            (ii) its r.m.s. value
            (iii) its instantaneous value when $t = 2\mu s$.
(b) Calculate the power developed in a 2.2 KΩ resistor when the above current flows through it.
(c) If the instantaneous value of voltage supplied to the circuit is $v = 100 \sin 314200\, t$ V draw a phasor diagram showing the r.m.s. values of voltage and current.

(Ans. (a) (i) 50 KHz (ii) 70.7 mA (iii) 35.8 mA (b) 11 W (c) I lags V by $\pi/12$).

6. By means of a phasor representation to a suitable scale, add the following voltages:

$$E_1 = 100 \sin \omega t, \ E_2 = 75 \sin \left(\omega t + \frac{\pi}{3} \right), \ E_3 = 60 \sin \left(\omega t - \frac{\pi}{2} \right).$$

Represent the resultant voltage in similar form.

(Ans. $E_R = 135 \sin \left(\omega t + 0.052^c \right)$.

7. A sinusoidal voltage has an r.m.s. value of $100 \sqrt{2}$ V at a frequency of 50 Hz. Commencing from zero in the cycle and rising positively, find the time taken for the voltage to reach a value of 100 V. (Ans. 1.67 ms).

8. An instantaneous value of current is given as $i = 200 \sqrt{2} \sin 877\, t$ mA. Calculate: (i) the r.m.s value of the current
            (ii) the frequency
            (iii) the instantaneous value of the current when t is 0.001 second.

(Ans. (i) 0.2 A (ii) 139.6 Hz (iii) 0.22 A).

9. A non-inductive resistor takes $10 \sqrt{2}$ A from a single phase 220 V mains, the voltage and current being sinusoidal. Plot graphs of current, voltage, and power over one complete cycle. Determine the average power dissipated (i) by calculation and (ii) from the waveform of power. (Ans. 3110.8 W).

10. Write down the mathematical expression representing a sinusoidal voltage having a peak value of 500 V and a periodic time of 10 ms and leading a current by a phase angle of $\frac{\pi}{10}$. $\left[\text{Ans. } 500 \left(\sin 200\pi t + \frac{\pi}{10} \right) \right].$

118

7. SINGLE PHASE A.C. CIRCUITS

7. *Uses a.c. circuit theory to solve simple series a.c. circuit problems.*
   a. *Draws the phasor diagrams and related voltage and current waveforms for circuit comprising.*

- ▶ *Pure resistance*
- ▶ *Pure inductance*
- ▶ *Pure capacitance*

PURE RESISTANCE

An a.c. voltage source is applied across a pure resistor R, where V is the r.m.s. voltage and $I = \dfrac{V}{R}$ r.m.s. current, Fig. 145.

Fig. 145 Pure non-inductive resistance.

WAVEFORMS

Fig. 146 shows the current in phase with the voltage.

Fig. 146 In phase waveforms

PHASORS

Fig. 147 shows the phasor diagram. If I is the reference phasor then V is in phase with I.

119

Fig. 147 Phasor diagram for a pure resistance.

PURE INDUCTANCE

An a.c. voltage source is applied across a pure inductance where V is the r.m.s. voltage and I is the r.m.s. current. Taking I as the reference phasor, the voltage leads the current by 90°. Fig. 148 shows the circuit and Fig. 149 the relevant waveforms. The phasor diagram is shown in Fig. 150.

Fig. 148 Pure inductance.

WAVEFORMS

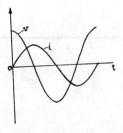

Fig. 149 Out of phase waveforms, i lags v.

PHASORS

Fig. 150 Phasor diagram for a pure inductance.

PURE CAPACITANCE

An a.c. voltage source is applied across a pure capacitor where V is the r.m.s. voltage and I is the r.m.s. current, Fig. 151 shows the circuit. Taking I as the reference phasor, the voltage lags the current by 90°.

Fig. 151 Pure capacitance.

The waveforms and phasors are shown in Fig. 152 and Fig. 153.

WAVEFORMS

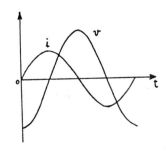

Fig. 152 Out of phase waveforms, i leads v

PHASORS

Fig. 153 Phasor diagram for a pure capacitance.

b. Describes inductive reactance and capacitive reactance in terms of impending the flow of an alternating current and use basic relationships to solve simple problems.

INDUCTIVE REACTANCE

For Fig. 148

$$X_L = \text{inductive reactance} = \frac{V}{I} \text{ (Ohm's Law)}$$

$$X_L = 2\pi f L \ (\Omega)$$

CAPACITIVE REACTANCE

For Fig. 151

$$X_c = \text{capacitive reactance} = \frac{V}{I} \text{ (Ohm's Law)}$$

$$X_c = \frac{1}{2\pi f C} \ (\Omega).$$

WORKED EXAMPLE 45

(a) A pure resistor is subjected to an a.c. source of 100 V r.m.s. and the current drawn is 10 A r.m.s. Calculate the resistance.

(b) A pure inductor is subjected to an a.c. source of 100 V r.m.s. (f = 50 Hz) and the current is 10 A r.m.s. Calculate the reactance and hence the inductance.

(c) A pure capacitor is subjected to an a.c. source of 100 V r.m.s. (f = 50 Hz) and the current is 10 A r.m.s. Calculate the reactance and hence the capacitance.

SOLUTION 45

(a) $R = \dfrac{V}{I} = \dfrac{100}{10} = 10 \ \Omega$

(b) $X_L = \dfrac{V}{I} = \dfrac{100}{10} = 10 \ \Omega$

      $X_L = 2\pi f L$

      $L = \dfrac{X_L}{2\pi f} = \dfrac{10}{2\pi 50} = 31.8 \ \text{mH}$

(c) $X_c = \dfrac{V}{I} = \dfrac{100}{10} = 10 \ \Omega$

      $X_c = \dfrac{1}{2\pi f C}$

      $C = \dfrac{1}{2\pi f X_c} = \dfrac{1}{2\pi 50 \, (10)} = 318.3 \ \mu\text{F}.$

WORKED EXAMPLE 46

(a)   An inductor of 50 mH when subjected to an a.c. supply of 50 Hz draws 100 mA current. Calculate the inductive reactance and the a.c. voltage.

(b)   A capacitor of 500 μF when subjected to an a.c. supply of 150 Hz draws 200 mA current. Calculate the capacitive reactance and the a.c. voltage.

SOLUTION 46

(a) $X_L = 2\pi f L = 2\pi \times 50 \times 50 \times 10^{-3} = 15.7 \ \Omega$

      $V = IX_L = 100 \times 10^{-3} \times 15.7 = 1.57 \ \text{V}$

(b) $X_c = \dfrac{1}{2\pi f C} = \dfrac{1}{2\pi \times 150 \times 500 \times 10^{-6}} = 2.12 \ \Omega$

      $V = IX_c = 200 \times 10^{-3} \times 2.12 = 0.424 \ \text{V}.$

c. Draws phasor diagrams corresponding to L-R and C-R series circuits.

L-R SERIES CIRCUIT

An inductor normally possesses an a.c. resistance R which is shown in this case as a resistor in series with a pure inductor. Fig. 154 shows the equivalent circuit of an inductor.

Fig. 154 L-R circuit

THE PHASOR

If I in the reference vector, V_R is in phase with I and V_L is leading I by 90°. Fig. 155 shows the phasor of an inductor.

Fig. 155 Phasor diagram of an inductor

The resultant vector is V as shown in Fig. 155. The voltage triangle is shown in Fig. 156.

Fig. 156 Voltage triangle.

124

and applying Pythagoras theorem

$$V^2 = V_L^2 + V_R^2$$

$$V_L = IX_L \text{ and } V_R = IR$$

$$V^2 = I^2X_L^2 + I^2R^2 = I^2\left(X_L^2 + R^2\right)$$

$$V = I\sqrt{X_L^2 + R^2}.$$

CR SERIES CIRCUIT

RC Circuit is shown in Fig. 157

Fig. 157 RC circuit.

Fig. 157 shows the r.m.s. current I flowing through R and C. I is the reference vector since it is common through R through C.

THE PHASOR

Fig. 158 shows the phasor diagram for the RC circuit in series. V_R is in phase with I and V_c lags I by 90°.

Fig. 158 The phasor diagram for a capacitor in series with a resistor.

125

The voltage triangle is shown in Fig. 159

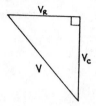

Fig. 159 Voltage triangle

$$V^2 = V_R^2 + V_C^2$$

$$V_C = IX_C \text{ and } V_R = IR$$

$$V^2 = I^2R^2 + I^2X_C^2 = I^2\left(R^2 + X_C^2\right)$$

$$V = I\sqrt{R^2 + X_C^2} \ .$$

d. Defines impedance as $Z = \dfrac{V}{I}.$

IMPEDANCE

For Fig. 154 and Fig. 157, impedance is defined as

$$Z = \frac{V}{I}.$$

For Fig. 154 $\quad Z = \dfrac{V}{I} = \dfrac{I\sqrt{X_L^2 + R^2}}{I}$

$$Z = \sqrt{R^2 + X_L^2}$$

For Fig. 157 $\quad Z = \dfrac{V}{I} = \dfrac{I\sqrt{R^2 + X_C^2}}{I}$

$$Z = \sqrt{R^2 + X_C^2} \ .$$

e. *Derives impedance triangle from voltage triangles and shows that* $Z^2 = R^2 + X^2$
and that $\tan \phi = \dfrac{X}{R}$, $\sin \phi = \dfrac{X}{Z}$ *and* $\cos \phi = \dfrac{R}{Z}$.

The impedance triangles are simply derived from the voltage triangle by dividing each side of the Fig. 156 and Fig. 159 by I since it is common.

Then we have Fig. 160 and Fig. 161 which are the impedance triangles.

Fig. 160 Impedance triangle for *LR* circuit

Fig. 161 Impedance triangle for *RC* circuit.

From the impedance triangles we get the following equations, by applying Pythagoras theorem.

$$Z^2 = R^2 + X_L^2 \text{ and } Z^2 = R^2 + X_C^2.$$

If ϕ is the phase angle, then we have from each triangle of Fig. 160 and Fig. 161

$\tan \phi = \dfrac{X_L}{R}$ $\qquad\qquad$ $\tan \phi = \dfrac{X_C}{R}$

$\cos \phi = \dfrac{R}{Z}$ $\qquad\qquad$ $\cos \phi = \dfrac{R}{Z}$

$\sin \phi = \dfrac{X_L}{Z}$ $\qquad\qquad$ $\sin \phi = \dfrac{X_C}{Z}$.

127

f. Applies equations in c, d, and e to the solution of single branch L-R and C-R series circuits at power and radio frequencies.

WORKED EXAMPLE 47

In an R-C series circuit the a.c. p.ds across R and C respectively 30 V and 40 V. Calculate the total a.c. p.d. applied.

SOLUTION 47

From the voltage triangle of Fig. 159

$$V^2 = V_R^2 + V_C^2$$

$$V = \sqrt{V_R^2 + V_C^2}$$

$V_R = 30$ V and $V_C = 40$ V

$$V = \sqrt{30^2 + 40^2} = \sqrt{900 + 1600}$$

$$= \sqrt{2500} = 50 \text{ V}$$

WORKED EXAMPLE 48

In an L-R series circuit the a.c. supply voltage is 240 V and that across R is 100 V. Calculate the p.d. across the inductor.

SOLUTION 48

From the voltage triangle of Fig. 156

$$V^2 = V_L^2 + V_R^2$$

$$240^2 = V_L^2 + 100^2$$

$$V_L^2 = 240^2 - 100^2 = 57600 - 10000$$
$$= 47600$$

$$V_L = 218.2 \text{ V.}$$

WORKED EXAMPLE 49

In the circuit shown in Fig. 162. Calculate   (i)   the supply current
                                                  (ii)   the supply phase angle
                                                  (iii)   the p.d.s across C and R.

Fig. 162 R-C series circuit

SOLUTION 49

(i) $Z = \sqrt{R^2 + X_C^2} = \sqrt{400^2 + 398^2} = 564.3 \ \Omega$

   where $X_C = \dfrac{1}{2\pi f C} = \dfrac{1}{2\pi \times 100 \times 4 \times 10^{-6}} = 398 \ \Omega$

   $I = \dfrac{V}{Z} = \dfrac{40}{564.3} = 0.071$ A.

(ii) $\tan \phi = \dfrac{X_C}{R} = \dfrac{398}{400} = 0.995$ from the impedance triangle of Fig. 161,

   $\phi = 44.9°$ is the phase angle

(iii) $V_C = IX_C = 0.071 \times 398 = 28.3$ V

   $V_R = IR = 0.071 \times 400 = 28.4$ V.

g. *Shows graphically that were currents and voltages are sinusoidal:*
   ▸ *for a purely resistive a.c. circuit, average power is VI*
   ▸ *for a purely reactive a.c. circuit, average power is zero*
   ▸ *for a resistive/reactive a.c. circuit, average power depends upon phase-angle.*

POWER IN A PURELY RESISTIVE A.C. CIRCUIT

At any instant, t, the power is iv, $p = iv$, i and v are in phase. Fig. 163 shows the circuit and the relevant waveforms.

Fig. 163 Circuit and i, v waveforms.

The current through R and the voltage across R are in phase as shown, the r.m.s. current is I and the r.m.s. voltages is V the average power $P = IV = \dfrac{V^2}{R} = I^2R$.

POWER IN A PURELY REACTIVE A.C. CIRCUIT

Fig. 164 and Fig. 165 show the circuits and relevant i, v waveforms.

Fig. 164 The pure inductor and i, v , waveforms.

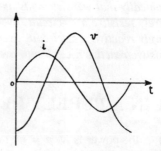

Fig. 165 The pure capacitor and i, v , waveforms

At any instant, t, the power is iv, $p = iv$, i and v are out of phase by 90° when $i = 0$, v_L is maximum when $v_L = 0$, i is maximum. Similarly for the capacitor when $i = 0$, v_C is negative maximum, when i is maximum v_C is zero and the average power is zero.

POWER IN A RESISTIVE/REACTIVE A.C. CIRCUIT

The average power in a resistive/reactive circuit depends on the phase angle.

$$P = IV \cos \phi.$$

Resistive/reactive a.c. circuits are shown in Fig. 166 and Fig. 167

Fig. 166 *LR* circuit

Fig 167 *CR* circuit

For a pure resistor $\phi = 0$ and $P = IV$.

For a pure capacitor $\phi = 90°$ and $P = IV \cos 90° = 0$.

For a pure inductor $\phi = 90°$ and $P = IV \cos 90° = 0$.

In an *LR* and *RC* circuits I and V normally have a phase angle ϕ and the Power is given by $IV \cos \phi$. Therefore the average power in a resistive/reactive circuit depends on the phase angle.

131

h. States that P = VI cos φ for sinusoidal waveforms.

If V leads I by an angle ϕ then $P = I(V\cos\phi) = VI\cos\phi$ active power where $V\cos\phi$ is the active component of voltage. Fig. 168 shows this situation. If I leads V by an angle ϕ, then $P = V(I\cos\phi) = VI\cos\phi$ active power where $I\cos\phi$ is the active component of current Fig. 169 shows this situation.

Fig. 168 V leads I by an angle ϕ.

Fig. 169 I leads V by an angle ϕ.

i. Derives the power triangle from the voltage triangle and identifies true power (P), apparent power (S), and reactive volt amperes (Q).

From the voltage triangles of Fig. 157 and Fig. 160, we multiply each side by I, we have the power triangles as shown in Fig. 170 and Fig. 171 respectively.

Fig. 170 Power triangle for inductive circuit.

132

Fig. 171 Power triangle for capacitive circuit.

Q = reactive volt amperes = $IV \sin \phi$
S = apparent power = IV
P = true power = $IV \cos \phi$.

j. Defines power factor as true power/apparent power and show that where V and I are sinusoidal, power factor = cos φ.

POWER FACTOR

Power factor $= \dfrac{\text{ture power}}{\text{apparent power}}$

$= \dfrac{IV \cos \phi}{IV}$

$= \cos \phi$.

k. Explains that power dissipation in series L-R and C-R a.c. circuits is I²R.

POWER DISSIPATION IN SERIES *L-R* A.C. CIRCUITS

The power dissipation in a pure inductor is zero and hence the only which is dissipated is in the resistor R. This power is $P = I^2 R$. Fig. 172 shows in L-R a.c. circuit.

Fig. 172 Power in *L-R* a.c. circuit.

133

POWER DISSIPATION IN SERIES
C-R A.C. CIRCUIT

Similarly the power dissipation in a pure capacitor is zero and the only power which is dissipated is in the resistor R. Again this power is $P = I^2 R$. Fig. 173 shows the *R-C* a.c. circuit.

Fig. 173 Power in *R-C* a.c. circuit.

l. Uses phasor diagrams to solve simple series L, C and R a.c. circuits.

R-L-C SERIES AC CIRCUIT

Fig. 174 shows the series *RLC* a.c. circuit subjected to an a.c. supply of r.m.s. voltage V and r.m.s. current I.

Fig. 174 *R-L-C* series circuit.

There are three cases to consider

a. If $V_L > V_C$

b. If $V_L < V_C$

c. If $V_L = V_C$.

The corresponding phasor diagrams are shown in Figs. 175, 176 and 177.

Fig. 175 Phasor for $V_L > V_C$

Fig. 176 Phasor for $V_L < V_C$

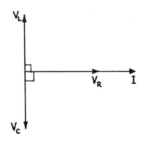

Fig. 177 Phasor for $V_L = V_C$

From these phasor diagrams, we have the impedance triangles shown in Fig. 178 and Fig. 179 corresponding to Fig. 175, Fig. 176, Fig. 177 has no impedance triangle, $Z = R$.

135

Fig. 178 Impedance triangle for the inductive case.

Fig. 179 Impedance triangle for the capacity case.

From Fig. 178 we have

$$Z = \sqrt{R^2 + (X_L - X_C)^2}.$$

From Fig. 179 we have

$$Z = \sqrt{R^2 + (X_C - X_L)^2}.$$

WORKED EXAMPLE 50

In the circuit shown in Fig. 180 calculate the following:
(i)   The total impedance and phase angle of the circuit.
(ii) The voltage across the capacitor.
(iii) The voltage across the resistor.
(iv) The voltage across the coil.
(v)   The power developed in the circuit.
(vi) The power factor of the circuit.

Fig. 180 A series *RLC* a.c. circuit.

SOLUTION 50

(i) $X_L = 2\pi f L = 2\pi \times 1000 \times 52 \times 10^{-3} = 326.7 \, \Omega$

$X_C = \dfrac{1}{2\pi f C} = \dfrac{1}{2\pi \times 1000 \times 15 \times 10^{-6}} = 10.6 \, \Omega.$

From the impedance triangle of Fig. 178

$Z = \sqrt{100^2 + (326.7 - 10.6)^2} = 331.5 \, \Omega$

using the equation $Z = \sqrt{R^2 + (X_L - X_C)^2}.$

(ii) $V_C = IX_C = 0.166 \times 10.6 = 1.76$ volts

where $I = \dfrac{V}{Z} = \dfrac{55}{331.5} = 0.166$ A.

(iii) $V_R = IR = 0.166 \times 100 = 16.6$ volts

(iv) $V_L = IX_L = 0.166 \times 326.7 = 54.23$ volts

(v) $P = I^2R = 0.166^2 \times 100 = 2.76$ watts

(vi) $\cos \phi = \dfrac{R}{Z} = \dfrac{100}{331.5} = 0.302.$

WORKED EXAMPLE 51

For the circuit shown in Fig. 181.

Fig. 181 Series *RLC* a.c. circuit.

137

Calculate: (i)   the impedance,

(ii) the total current,

(iii) the voltage across each component,

(iv) the power factor,

(v)   the power dissipated.

SOLUTION 51

(i) $X_L = 2\pi f L = 2\pi \times 2000 \times 2.5 \times 10^{-3} = 31.42\ \Omega$

$$X_C = \frac{1}{2\pi f C} = \frac{1}{2\pi \times 2000 \times 8 \times 10^{-6}} = 9.95\ \Omega$$

$$Z = \sqrt{R^2 + (X_L - X_C)^2} = \sqrt{15^2 + (31.42 - 9.95)^2} = 26.2\ \Omega$$

(ii) $I = \dfrac{V}{Z} = \dfrac{2.5}{26.2} = 95.4\ \text{mA}$

(iii) $V_R = 95.4 \times 10^{-3} \times 15 = 1.43\ \text{V}$

$V_L = 95.4 \times 10^{-3} \times 31.42 = 2.997\ \text{V} \approx 3\ \text{V}$

$V_C = 95.4 \times 10^{-3} \times 9.95 = 0.949\ \text{V}$

(iv) $\cos \phi = \dfrac{R}{Z} = \dfrac{15}{26.2} = 0.573$

(v) $P = I^2 R = 0.0954^2 \times 15 = 0.137\ \text{W}.$

_____

m.   *Defines series resonance as occuring when the supply voltage and current are*
in phase and sketches a phasor diagram showing that:
* $V_L = V_c$ *at resonance*
* V_L *and* V_c *may be much greater than the supply voltage.*

SERIES RESONANCE

Series resonance occurs when the supply voltage and current are in phase.
Fig. 182 shows this condition.

138

Fig. 182 Series resonance.

At resonance

$V_L = V_C$ and the supply voltage $V = V_R$

$$I_o 2\pi f_o L = I_o \frac{1}{2\pi f_o C} \quad \ldots (1)$$

where I_o is the current in the circuit at resonance which is maximum and the frequency f_o is the resonant frequency from equation (1) we have

$$f_o = \frac{1}{2\pi \sqrt{LC}}$$

$$V_L = I_o X_L = \frac{V}{R} X_L = V \left(\frac{X_L}{R} \right)$$

$$V_c = I_o X_C = \frac{V}{R} X_C = V \left(\frac{X_C}{R} \right) .$$

The ratio $\dfrac{X_L}{R}$ or $\dfrac{X_C}{R}$ may be much greater than unit, and hence the p.d. across X_L or X_C is much greater than the supply voltage, V.

$$Q = \text{Quality factor} = \frac{X_L}{R} = \frac{X_C}{R} .$$

WORKED EXAMPLE 52

A series circuit consists of a non-inductive resistance of 20 Ω, a capacitance of 10 μF, and pure inductance of 150 mH. The a.c. supply voltage is 200 V at 100 Hz.

Calculate (i)   the circuit impedance
           (ii)   the current in the circuit
           (iii) the phase angle.

If another component is connected in series with the above in order that the impedance of the circuit is only 20 Ω or the phase angle is zero or the current in the circuit is maximum. Determine the new value of current.

SOLUTION 52

Fig. 183 shows the series *LRC* a.c. circuit.

Fig. 183 *LRC* series circuit.

(i) $X_L = 2\pi fL = 2\pi \times 100 \times 150 \times 10^{-3} = 94.25\ \Omega$

$X_L = 94.25\ \Omega$

$X_C = \dfrac{1}{2\pi fC} = \dfrac{1}{2\pi \times 100 \times 10 \times 10^{-6}} = 159.2\ \Omega$

$X_C = 159.2\ \Omega$

$Z = \sqrt{20^2 + \left(159.2 - 94.25\right)^2} = 67.96\ \Omega$

(ii) $I = \dfrac{V}{Z} = \dfrac{200}{67.96} = 2.94\ \text{A}$

(iii) $\tan \phi = \dfrac{X_C - X_L}{R} = \dfrac{159.2 - 94.25}{20} = 3.25$

$\phi = 72.9°.$

The other component to be added must be a pure coil of reactance
159.2 – 94.25 = 65 approximately

$X_L = 2\pi fL = 65$

$L = \dfrac{65}{2\pi \times 100} = 103\ \text{mH}$

140

such that $X_L = X_C$

$$I_o = \frac{V}{R} = \frac{200}{20} = 10 \text{ A}.$$

This problem is only a theoretical exercise since there is no pure coil in practice. Fig. 184 shows the new circuit for resonance.

Fig. 184 103 mH is added to cause resonance.

WORKED EXAMPLE 53

A series circuit consists of a coil of inductance of 250 mH, a resistor of resistance 30 Ω, and a capacitor of capacitance C. The circuit resonates when the supply is 240 V and the power frequency is 50 Hz.

Calculate: (i)   the value of C
(ii)   the resonant current
(iii)   the voltage across the inductor
(iv)   the power dissipated
(v)   the Q-factor of the circuit where $Q = \dfrac{2\pi f_o L}{R}$.

SOLUTION 53

(i) $X_L = 2\pi f_o L = 2\pi \times 50 \times 250 \times 10^{-3} = 78.54 \ \Omega$

$X_L = 78.54 \ \Omega$

$X_C = 78.54 = \dfrac{1}{2\pi f_o C}$

$C = \dfrac{1}{2\pi \times 50 \times 78.54} = 40.5 \ \mu\text{F}$

141

(ii) $I_o = \dfrac{V}{R} = \dfrac{240}{30} = 8 \text{ A}$

(iii) $V_L = I_o X_L = 8 \times 78.54 = 628.32 \text{ volts}$

(iv) $P = I_o^2 R = 8^2 \times 30 = 64 \times 30 = 1920 \text{ W}$

(v) $Q = \text{quality factor} = \dfrac{\omega_o L}{R} = \dfrac{78.54}{30} = 2.618$

$Q = \dfrac{628.32}{240} = \dfrac{V_L}{V} = 2.618$

which verifies that V_L is equal to Q times the supply voltage V.

SUMMARY 7

PURE RESISTOR $R = \dfrac{V}{I}$ (Ω) ohm's law

PURE INDUCTOR $X_L = \dfrac{V}{I}$ (Ω) ohm's law

PURE CAPACITOR $X_C = \dfrac{V}{I}$ (Ω) ohm's law

INDUCTIVE REACTANCE $X_L = 2\pi f L$ (Ω)

Fig. 185

Fig. 186

142

CAPACITIVE REACTANCE $X_C = \dfrac{1}{2\pi f C}$ (Ω)

SERIES INDUCTIVE *RL* CIRCUIT $V^2 = V_R^2 + V_L^2$

SERIES CAPACITIVE *RC* CIRCUIT $V^2 = V_R^2 + V_C^2$

IMPEDANCES $\quad Z = \sqrt{R^2 + X_L^2}$

$$Z = \sqrt{R^2 + X_C^2}$$

IMPEDANCE TRIANGLE $\tan\phi = \dfrac{X}{R}$

$$\sin\phi = \dfrac{X}{Z}$$

$$\cos\phi = \dfrac{R}{Z}$$

POWER TRIANGLE Q = REACTIVE VOLT AMPERES = $IV \sin\phi$

S = APPARENT POWER = IV

P = TRUE POWER = $IV \cos\phi$

RESISTIVE POWER = $P = I^2 R = IV = \dfrac{V^2}{R}$

RLC CIRCUIT $Z = \sqrt{R^2 + \left(X_L - X_C\right)^2}$

RESONANCE: $Z = R \qquad f_o = \dfrac{1}{2\pi\sqrt{LC}}$

$$I_o = \dfrac{V}{R}$$

$$V_L = QV \qquad V_C = QV$$

$$Q = \dfrac{X_L}{R} \qquad Q = \dfrac{X_C}{R}.$$

Fig. 186

EXERCISES 7

1. A capacitor of 100 μF capacitance is subjected to an a.c. supply of frequency 100 Hz. Determine the capacitive reactance. (Ans. 15.9 Ω).

2. An inductor of 50 mH inductance is subjected to an a.c. supply of frequency 60 Hz. Determine the inductive reactance. (Ans. 18.9 Ω).

3. The reactance of a capacitor is 100 Ω when subjected to an a.c. supply of 50 V. Calculate the current through the circuit. (Ans. 0.5 A).

4. The inductive reactance of a coil is 5 Ω when it is subjected to an a.c. supply which draws 5 A current. Calculate the a.c. supply. (Ans. 25 V).

5. Sketch neatly a sine wave of current, and in relation to this current sketch the voltage waveform you would expect when this current passes through (i) a resistance (ii) an inductance (iii) a capacitance. Label your diagrams.

6. A series circuit consists of an inductor of inductance 0.5 H and a resistor of resistance 50 Ω. A 240 V a.c. supply of frequency of 50 Hz is connected across the above circuit.

   Calculate (i) the impedance of the circuit
   (ii) the current drawn from the a.c. supply
   (iii) the p.d. across the pure inductor
   (iv) the p.d. across the resistor and hence check that the supply voltage is 240 volts.

   (Ans. (i) 164.9 Ω (ii) 1.46 A (iii) 229.3 V (iv) 73 V).

7. A coil takes a current of 5 A when connected to a 100 V d.c. supply, when the same coil is connected to a 100 V 50 Hz a.c. supply the current drawn is 1 A. Explain why these currents differ.

   Calculate: (i) the resistance of the coil
   (ii) the impedance of the coil at 50 Hz.

   (Ans. 20 Ω, 100 Ω).

8. An inductor takes 10 A and dissipates 500 W when connected to a 200 V 50 Hz a.c. supply.

   Calculate (i) the impedance,
   (ii) the power factor,
   (iii) the resistance,
   (iv) the inductance.

   (Ans. (i) 20 Ω (ii) 0.25 (iii) 5 Ω (iv) 61.6 mH).

9. The power taken by an inductive circuit when connected to a 240 V, 50 Hz supply is 1000 watts, and the current is 16 A.

Calculate (i) the power factor of the circuit,
(ii) the impedance,
(iii) the resistance,
(iv) the reactance,
(v) the inductance.

(Ans. (i) 0.26 (ii) 15 Ω (iii) 3.9 Ω (iv) 14.48 Ω (v) 46.1 mH).

10. An *RLC* series circuit has a resistance of 15 Ω, an inductance of 300 mH and a capacitance of 100 μF.

Calculate (i) the resonant frequency,
(ii) the supply current,
(iii) the p.d. across the inductance,
(iv) the p.d. across the capacitance,
(v) the power at resonance.
Given that the applied a.c. voltage is 240 V.

(Ans. (i) 29.06 Hz (ii) 16 A (iii) 876.4 V (iv) 876.4 V (v) 3840 W).

11. A variable inductor with negligible effective resistance is connected in series with a resistor and a capacitor. The whole circuit is applied across an a.c. supply of 240 V, 50 Hz, the inductor was adjusted to give a maximum current of 1 A.

The p.d. across the capacitor is 500 volts at this maximum current. Calculate the value of the resistance, the capacitance and inductance.

(Ans. 240 Ω, 6.37 μF and 1.59 H).

12. A series circuit consists of a coil of resistance 50 Ω and inductance 0.05 H, and a capacitor 0.2 μF. The circuit is subjected to an a.c. supply of 300 V.

Calculate (i) the resonant frequency,
(ii) the resonant current,
(iii) the p.d. across the capacitor,
(iv) the quality factor, Q.

(Ans. (i) 1591.6 Hz (ii) 6 A (iii) 3000 V (iv) 10).

13. An inductive circuit of resistance 10 Ω and inductance 50 mH is connected in series with a capacitor. When connected across a 240 V 50 Hz supply, the current taken is 5 A leading. What is the value of the capacitor? Repeat the calculation if the current taken is 2 A lagging.

(Ans. 51 μF, 30.6 μF).

145

8. MEASURING INSTRUMENTS AND MEASUREMENT

8. Understands the operation and limitations of measuring instruments and uses them correctly in a wide range of basic applications.

a. Describes with the aid of given diagrams, the principles of operation of moving coil instruments and explains the need for shunts and multipliers to extend the range of basic electrical indicating instruments.

MOVING COIL INSTRUMENT

This instrument is basically a coil of fine insulated copper wire of N turns wound on a rectangular former as shown in Fig. 187.

Fig. 187 Coil of N turns wound on a rectangular former.

This rotates about a central spindle in a strong magnetic field. The currents flowing through the narrower sides are parallel with the field B and the forces exerted are zero. The current flowing through the longer sides are at right angles and therefore the forces are maximum and equal to $F = BIl$ where l = length of longer sides.

A soft iron cylinder is place as shown inf Fig. 188 and together with a permanent magnet produces a radial field as shown in Fig. 189.

Fig. 188 Rectangular coil on soft iron cylinder.

146

Fig. 189 Rectangular coil placed in a permanent magnet.

A pointer is attached at right angles to the spindle.

The coil is pivoted on jewelled bearings and its rotation is resisted by hair springs above and below the coil.

Each of the vertical conductors (long sides) of the coil is placed at right angles to the radial field produced by the permanent magnets.

If the Left Hand Rule is applied, the following situations occur which are shown in details in Figs. 190, 191, 192.

Fig. 190 One turn of the rectangular coil

Fig. 191   Force going into the paper for the left hand
           side of the rectangular coil.

147

Fig. 192   Force coming out of the paper for the right hand side
of the rectangular coil.

A couple of forces occur as in Fig. 193. and the coil rotates about the central spindle.

Fig. 193 Couple of forces $F \times b$

CURRENT SENSITIVITY

Couple of forces is defined as $F \times b = BIl \times N \times b = c\,\theta$ where c is the controlling constant and θ is the deflection of the pointer in radians

$$\theta c = BIlNb$$

$$\frac{\theta}{I} = \frac{BNA}{c} \text{ where } A = bl \text{ the area of the rectangle}$$

$\dfrac{\theta}{I}$ = deflection per unit current = current sensitivity.

For maximum current sensitivity, B should be large, N is large, and A is large and c must be as small as it is possible.

THE SHUNT

A galvanometer is a sensitive ammeter, measuring very small currents. In order to obtain larger currents, a resistor called the *shunt* is connected in parallel with the galvanometer.

Fig. 194 The shunt.

$$I_{sh} = \frac{R_G}{R_G + R_{sh}} I$$

$$I_G = \frac{R_{sh}}{R_G + R_{sh}} I$$

$$I = I_G + I_{sh} = I_G + I_G \frac{R_G}{R_{sh}} = I_G \left(\frac{R_{sh} + R_G}{R_{sh}} \right)$$

$$\frac{I}{I_G} = \frac{R_{sh}}{R_{sh}} + \frac{R_G}{R_{sh}} = 1 + \frac{R_G}{R_{sh}} \quad \text{or} \quad \frac{R_{sh}}{R_G} = \frac{1}{\dfrac{I}{I_G} - 1}$$

$$R_{sh} = \frac{R_G}{\left(\dfrac{I - I_G}{I_G} \right)} \ .$$

The terminals of the extended scale instrument, are, A' and B'.

THE MULTIPLIER

A multiplier is a large resistance R_M connected in series with the galvanometer in order to extend the voltage scale. Fig. 195 shows the principle of multiplier.

Fig. 195 The multiplier.

From Fig. 195 we have

$$V = I_G(R_G + R_M)$$

re-arranging the formula for R_M.

$$R_M = \frac{V}{I_G} - R_G.$$

The new voltmeter terminals are now A and B.

WORKED EXAMPLE 54

A moving coil instrument has a resistance of 15 Ω and the full scale deflection is 50 mA. How would you adapt this instrument for use as an ammeter reading 0-5 amperes?

SOLUTION 54

In order to adapt the moving coil instrument to read full scale deflection of 5 A, a shunt is connected across the instrument as shown in Fig. 196.

Fig. 196 Extended scale of an ammeter.

$$I_{sh} = \frac{15}{15 + R_{sh}} 5 = 4.95$$

$$15 + R_{sh} = \frac{15 \times 5}{4.95} = 15.151515$$

$$R_{sh} = 0.151515 \ \Omega.$$

WORKED EXAMPLE 55

A moving coil instrument has a resistance of 15 Ω and the full scale deflection is 50 mA. How would you adapt this instrument for use as a voltmeter reading 0-10 V?

SOLUTION 55

In order to adapt the moving coil instrument to read full scale deflection of 10 V, a multiplier resistor is connected in series as shown in Fig. 197.

From Fig. 197 we have

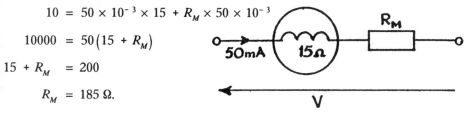

$$10 = 50 \times 10^{-3} \times 15 + R_M \times 50 \times 10^{-3}$$

$$10000 = 50\left(15 + R_M\right)$$

$$15 + R_M = 200$$

$$R_M = 185\ \Omega.$$

b. Uses ammeters and voltmeters correctly in d.c. circuit measurement.

A BASIC AMMETER

A moving coil instrument is basically a d.c. ammeter and is used to detect current from a few mA (milliamperes) to several hundred amperes. Galvanometers are used to measure small currents, only in the range of μA (micro-amperes). They have a large number of turns, are very sensitive and delicate instruments. Galvanometers have large resistances in the order of 50 Ω or 100 Ω but in conjunction with shunts, the resistance is greatly reduced. The higher the full scale deflection (f.s.d.) the smaller is the resistance between the terminals. Therefore, ammeters normally have a *very small resistance* between their terminals.

Ammeters are connected in series with the component whose current is required to be measured.

A BASIC VOLTMETER

A d.c. voltmeter is normally a moving coil d.c. ammeter with a large resistor connected in series and thus voltmeters normally have a very large resistance between their terminals.

Voltmeters are connected in parallel with the component whose voltage is required to be measured.

AMMETER AND VOLTMETER METHOD FOR MEASURING RESISTANCE

There are two cases to consider:
(a)   the measurement of a low value of resistance of the order of 10 Ω and
(b)   the measurement of a high value of resistance of the order of 5000 Ω.

Fig. 198 shows the connection of the ammeter for low values of resistance.

Fig. 198 Measurement of low values of resistance.

The current indicated on the ammeter is the total current of the circuit.

If the resistance of the voltmeter, R_v = 1500 Ω, then the currents I_R and I_v are given by the current divider as

$$I_R = \frac{1500}{1510} I = 0.9934I \quad \text{... (1)}$$

$$I_v = \frac{10}{1510} I = 0.0066I \quad \text{... (2)}$$

In order to measure the resistance, R, a set of readings of I and V are noted as the rheostat is varied, R is then calculated from the average value of the ratios of $\frac{V}{I}$, which will be slightly less than the actual value.

The percentage error in this case will be about 0.66. This is shown in equation (2) and I_R is taken to be approximately equal to the total current I.

Fig. 199 shows the connection of the ammeter for high values of resistance.

Fig. 199 Measurement of high values of resistance.

The current indicated on the ammeter is the actual current that flows through the resistance which is to be measured. But the voltmeter reading indicated will be slightly higher, since it will be measuring the p.d. of the ammeter as well, so in this case the value of resistance R will be slightly higher by considering the ratio of $\dfrac{V}{I}$.

In both the above methods, however, there are small errors.

c. Describes with the aid of diagrams the principles of operation of an ohm meter and uses an ohm meter for the measurement of resistance.

THE BASIC PRINCIPLE OF AN OHM METER

A moving coil is connected in series with a variable resistor R, a battery of 3 V and the two terminals A and B as shown in Fig. 200. When A and B are short circuited and R is adjusted for full scale deflection of current, then this corresponds to zero resistance between A and B.

The scale of ohmmeter is calibrated as shown in Fig. 201.

Fig. 200 The ohmmeter.

Fig. 201 Calibrated scale of an ohmmeter is non linear.

The short circuit is now replace with standard resistors of different values such as 100 Ω, 500 Ω, 1000 Ω, 10000 Ω and they are placed in turn and then the resistance scale is calibrated as shown in Fig. 201. Note that the scale will be non-linear.

To measure an unknown resistance R_x, it is placed between A and B terminals thus giving a deflection which gives its value.

USE OF AN OHMMETER

Short circuit A and B and adjust R to give zero resistance. Zero (corresponds to f.s.d. current). Then replace the short circuit with the unknown resistance and read off its value from the calibrated scale.

d. Explains the need for rectifier instruments and states the frequency and waveform limitation inherent in moving-coil rectifier instruments.

RECTIFIER INSTRUMENT

A moving-coil instrument is used in conjunction with a rectifier bridge circuit in order to provide an indication of alternating quantities. This arrangement is shown in Fig. 202.

Fig. 202 Rectifier instrument.

When A is positive with respect to B, current flows through D_1, through the moving coil meter, and D_2.

When B is positive with respect to A, current flows through D_3, the moving coil meter and D_4. The d.c. current, $I_{d.c.}$ flowing through the M.C., and the a.c. current I are connected by expression

$$I_{d.c.} = \frac{I}{1.11}$$

where 1.11 as the form factor of the sinusoidal waveform applied as the a.c. signal.

Rectifier instruments have their scales calibrated in terms of r.m.s. quantities and the a.c. signals are sinusoidal.

These instruments may be frequency limited. Rectifier moving coil instruments can be used on frequencies up to several KHz.

e. Use electronic and moving - coil multimeter correctly for the measurements of I and V in d.c. and a.c. circuits.

We have seen previously the voltmeters possess much higher resistances between their terminals, than ammeters.

154

voltmeters are therefore graded according to their 'resistance per volt' at full scale deflection (f.s.d.). The total resistance of the meter is that of the coil and the resistance of the multiplier. If the full scale deflection is 100 μA and the p.d. is 1 V between the terminals then the total resistance, R, is given by

$$R = \frac{1}{100 \times 10^{-6}} = \frac{10^6}{100} = 10000 \ \Omega$$

and the *resistance per volt* is 10000 Ω/V. A good grade voltmeter must have a resistance of at least 1000 Ω/V.

Electronic voltmeters are widely used where the resistance between the terminals is of the order of 10 MΩ and thus the current through the meter is negligible. Thus for accurate measurements electronic voltmeters are preferred for d.c. and a.c. circuit.

. Uses a wattmeter

THE DYNAMOMETER WATTMETER

The dynamometer wattmeter is the most commonly used instrument in a.c. circuits. The dynamometer instrument is a moving coil instrument in which the magnetic field is not provided by a permanent magnet, but by another pair of coils. The dynamometer principle is basically made up of two fixed field coils and a moving coil, the current passes through the fixed coils and sets up a magnetic field which is effectively uniform in the region around the moving coil. This is illustrated in the Fig. 203.

Fig. 203

The magnetic field is produced by the current I_1 flowing in two fixed circular coils which are connected in series in this case (it can be connected in parallel also). The field between the two coils is substantially uniform, and the moving coil also circular, and carrying current I_2 rotates in this field.

Therefore when a dynamometer instrument is used as a wattmeter the currents in the fixed and moving coils are different.

155

The fixed coils carry the load current, and the moving coil carries a current proportional to the voltage across the load.

The m.m.f.'s F_1 and F_2 act in the directions indicated, and the operating torque T is approximately given by

$$T = KF_1F_2 \sin \theta$$

since F_1 is proportional to I_1 and F_2 to I_2, and sin θ in numerically equal to sin θ

$$T = KI_1I_2 \sin \theta$$

when the instrument is used as an ammeter the moving coil is connected either in series or in parallel with the fixed coils, I_1 and I_2 are then both proportional to the current I being measured, so the torque is proportional to I^2, and the instrument reads true r.m.s.

The dynamometer instrument can thus be used on alternating current circuits. The wattmeter connection on a single-phase circuit are usually made as shown in Fig. 204.

Fig. 204   Wattmeter connection for measurement of power on a
single-phase system.   Normal connection.

Using these connections, the current coils measure the true current taken by the load, but the voltage coils measure the p.d. across the current coil in addition to the p.d. across the load. The result is that the power loss in the current coils of the instrument is included in the indication.   The error is usually small, but a compensation can be made if the resistance r_c of the current coils is known.   The power loss $I^2 r_c$ is then calculated and subtracted from the wattmeter reading.

Alternative connections are shown in Fig. 205.

Fig. 205   Wattmeter connections for measurement of power on a
single-phase system.   Alternative connections.

156

The voltage coil then measures the voltage across the load, but the current coil measures the current taken by the voltage coil in addition to the current taken by the load. The result is that the power loss in the voltage coil is indicated in the reading. Again the error is usually small, but compensation can be made if the resistance r_v

of the voltage coil is known. The power loss $\dfrac{V^2}{r_v}$ can then be calculated and subtracted from the wattmeter reading.

g. Uses double beam CRO in direct and indirect measurement of a.c. and d.c. voltages and to measure period and frequency.

CATHODE RAY OSCILLOSCOPE (C.R.O.)

A Cathode Ray Oscilloscope (C.R.O.) with two amplifiers (double beam) is used to measure voltages visually. That is, two waveforms may be displayed on the C.R.O. at any one time and thus the alternating voltage waveforms are studied. A special electronic circuit is built within the C.R.O. and is called *the time base*. The time base circuit produces an output voltage with saw-tooth waveform as shown in Fig. 206.

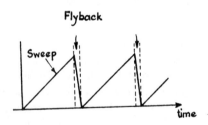

Fig. 206 Time base waveform.

The time base period = the sweeping time + the flyback time. This voltage is connected across the x-plates in such a way that the spot is scanned across the screen from left to the right relatively slowly and is then flies back to the left hand side very quickly. This cycle then is repeated. The voltage to be examined is connected across the y-plates.

If the frequency of the time base is adjusted to be the same as the frequency on the y-plates, a steady waveform will be displayed on the screen.

THE CATHODE RAY TUBE (C.R.T.)

The cathode ray tube is the heart of the C.R.O. and it is simply described here.

Fig. 207 Construction of an Electrostatic Cathode Ray Tube (C.R.T.).

The C.R.T. is used to study the waveforms of d.c. and a.c. electrical voltages and currents. The construction of a typical C.R.T. with electrostatic deflection and focusing is shown in Fig. 207. It consists basically of the electron gun, the deflecting system of plates and the fluorescent screen.

An indirectly heated cathode in the form of a nickel cylinder with an oxide coated end cap.   A cylindrical grid, houses the cathode with negative potential for controlling the emitted electrons from the cathode. The anodes A_1 and A_2 are discs with central, holes through which the electrons pass, a cylindrical anode, the focusing anode, is placed between A_1 and A_2. The beam passes between two pairs of parallel deflecting plates, which are mutually perpendicular, the y-plates and the x-plates.   A p.d. across the x-plates produces a horizontal deflection and a p.d. across the y-plates produces vertical deflection of the spot on the screen.

The funnel-shaped part of the C.R.T. is coated internally with graphite and is earthed. This allows the electrons which strike the screen to leak away. The screen however is coated with a fluorescent powder, such as zinc sulphide (blue glow) or zinc orthosilicate (blue-green glow).

The C.R.O. has a high input impedance which means that it does not draw a high current.   The C.R.O. is a visual electronic voltmeter measuring a.c. and d.c. waveforms.

To measure a voltage, the peak-to-peak distance of the trace is measured in centimetres using the graticule mounted in front of the screen.   Knowing the sensitivity in volts per centimetre then enables us to convert this reading into volts. The amplifier controls are calibrated in V/cm or mV/cm, these calibrations are approximate.

PERIOD AND FREQUENCY MEASUREMENT
X AND Y CONTROLS

The time base switch can be adjusted to measure the period and hence the frequency of the waveform. Let us assume that one complete cycle of a sinusoidal waveform is displayed on the screen, and that the length of the horizontal distance is 3.6 cm an the time base control is set as 100 μs/cm. The periodic time of the waveform is given by 3.6 cm $\times$ 100 μs/cm = 360 μs (provided the C.R.O. is calibrated).

$$T = 360 \times 10^{-6} \text{ s.}$$

The frequency of the waveform is found to be

$$f = \frac{1}{T} = \frac{1}{360 \times 10^{-6}} \text{ Hz}$$

$$= 2778 \text{ Hz}$$

$$f = 2.78 \text{ KHz}$$

Fig. 208 Period and frequency measurement of a sinusoidal waveform.

WORKED EXAMPLE 56

A sinusoidal waveform is displayed on the screen of an oscilloscope and the following measurements where made on the graticule of the screen:

Peak-to-peak distance = 5.7 cm (vertical) horizontal distance between the two peaks = 4.8 cm. The Y amplifier setting is 500 mV/cm and the time base setting is 100 μs/cm.

Determine the approximate r.m.s. voltage, period and frequency.

SOLUTION 56

Peak-to-peak voltage = 5.7 cm $\times$ 500 $\times$ 10^{-3} V/cm
= 2.85 V

Peak voltage = 1.43 V

R.M.S. voltage = $\dfrac{1.43}{\sqrt{2}}$ = 1 V approximately

Periodic time = 4.8 cm × 100 × 10^{-6} s/cm = 0.48 ms

Frequency = $\dfrac{1}{T}$ = $\dfrac{1}{0.48 \times 10^{-3}}$ = 2083 Hz approximately

LISSAJOU'S FIGURES

A more accurate way of measuring frequency is to use the C.R.O. to make a comparison between the unknown frequency and a signal of known frequency and a signal generator (oscillator). The unknown signal is applied to the Y plates and the output from the signal generator to the X plates, the time base being switched off. When the frequency of the signal generator is suitably adjusted, Lissajou's figures can be formed on the C.R.O. screen. The shape of the figure depends on the ratio of the two frequencies. For simple ratios such as 1:1 and 2:1 the figures are an ellipse and a figure of eight respectively.

h. Describes the principle of null method of measurement.

A sensitive ammeter or galvanometer is used with a centre zero (cz) as shown in Fig. 209.

Fig. 209 A centre zero galvanometer instrument.

The pointer is at the centre of the instrument and may deflect to the left or to the right from the centre, thus indicating that the current either flows from left to right or from right to left. A null method is a method which indicates that the current through the instrument is zero so that the p.d. across the terminals is zero, and thus the power loss in the instrument is negligible.

i. Describes with the aid of diagram the principle of the Wheatstone bridge and uses a Wheatstone bridge to measure resistance over a wide range.

WHEATSTONE BRIDGE

Four resistors are placed as shown in Fig. 210 forming a bridge.

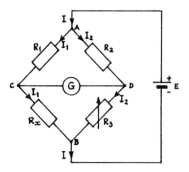

Fig. 210 Wheatstone bridge.

The d.c. supply is connected across one diagonal and the sensitive ammeter or galvanometer across the other diagonal. R_1 and R_2 are two standard resistors of high accuracy, R_3 is a four or five or six dial standard resistance box and R_x is the unknown resistance which is to be measured by means of a null method. By adjusting R_3 it can be arranged that no current flows through G.

THE BALANCED CONDITION

Let I be the total current drawn from the d.c. supply E, and let I_1 and I_2 be the current that will flow through R_1 and R_x, and R_2 and R_3 respectively. By adjusting R_3 so that no current flows through G, the pointer is at the centre of the instrument. The p.d. across CD is zero and the potential at C is equal to the potential at D.

p.d. across AC = p.d. across AD

$$I_1R_1 = I_2R_2 \quad \dots (1)$$

p.d. across CB = p.d. across BD

$$I_1R_x = I_2R_3 \quad \dots (2)$$

Dividing equations (1) and (2)

$$\frac{I_1R_1}{I_1R_x} = \frac{I_2R_2}{I_2R_3}$$

161

$$\frac{R_1}{R_x} = \frac{R_2}{R_3} \quad \dots \text{ (3)}$$

transposing equation (3)

$$R_x = \left[\frac{R_1}{R_2} \right] R_3$$

$\dfrac{R_1}{R_2}$ is called the *arm's ratio*

R_3 is multiplied by R_1 (R_1 is the *multiply*)

R_3 is divided by R_2 (R_2 is the *divide*)

R_1 can be one of three resistors. $R_1 = 10 \, \Omega$, $R_1 = 100 \, \Omega$, $R_1 = 1000 \, \Omega$.

R_2 can be one of three resistors $R_2 = 10 \, \Omega$, $R_2 = 100 \, \Omega$, $R_2 = 1000 \, \Omega$.

R_3 is a four, a five or a six dial variable resistance box.

UNITS	TENS	HUNDREDS	THOUSANDS	TEN THOUSANDS
$\times 1 \, \Omega$	$\times 10 \, \Omega$	$\times 100 \, \Omega$	$\times 1000 \, \Omega$	$\times 10000 \, \Omega$

HUNDRED THOUSANDS
$\times 100000 \, \Omega$

The basic principle of the Wheatstone bridge depends on the balanced equation

$$R_x = \left[\frac{R_1}{R_2} \right] R_3 .$$

For a resistance of 1.23 Ω

$$R_x = \left[\frac{10}{1000} \right] 123 = 1.23 \, \Omega$$

and for a resistance of 100000000 Ω

$$R_x = \left[\frac{1000}{10} \right] 1000000 = 100000000 \, \Omega$$

and thus the wheatstone bridge can accurately measure resistances over a wide range.

EXERCISES 8

1. Sketch the basic construction of a moving coil instrument, and explain the principle of operation.

2. A moving-coil milliammeter has a coil resistance of 10 Ω and gives a full scale deflection of current of 15 mA.

    (i) Calculate the value of the resistor required to enable the instrument to read 0-5 A.

    (ii) Calculate the value of the resistor required to enable the milliammeter to read 0-500 V.

    (Ans. (i) 0.03 Ω (ii) 33323.33 Ω).

3. A moving coil meter movement has a coil resistance of 400 Ω and a resistor of 600 Ω is connected in series with it. The movement requires a current of 2 mA for f.s.d. (full scale deflection). Calculate the shunt resistance required to change the movement into a 0-1 A ammeter.

    (Ans. 2.004 Ω).

4. Explain, with the aid of a suitable diagram the principle of operation of the Wheatstone bridge. Derive from first principles an expression giving, for the balanced condition, the unknown resistance in terms of the known quantities.

5. A 0 to 50 V meter has resistance of 500000 Ω what additional voltage multiplier resistor is required to enable the meter to indicate 0 to 500 V.

    (Ans. 4.5 MΩ).

6. For the Wheatstone bridge shown in Fig. 211 and it is at balance.

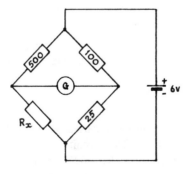

Fig. 211 Wheatstone bridge

163

Determine: (i) the value of R_x

(ii) the p.d. across the 100 Ω

(iii) the current through the 500 Ω resistor.

(Ans. (i) 125 Ω (ii) 4.8 V (iii) 9.6 mA).

7. Explain with the aid of a circuit diagram what is meant by a potential divider and show how two such potential dividers make up a wheatstone bridge circuit.

8. Compare the Wheatstone bridge with the ammeter and voltmeter methods for measuring resistance.

9. An ammeter of resistance 0.05 Ω has a shunt of resistance 0.015 Ω connected across it. What is the total current flowing when the current through the ammeter is 5 A?

(Ans. 21.7 A).

10. A voltmeter with resistance of 1000 Ω is used to measure the p.d. between points A and B of the circuit shown in Fig. 212. What is the voltmeter reading and how does this compare with the true value.

Fig. 212 Voltmeter measurement.

(Ans. 66.7 V).

11. An ammeter and a voltmeter are used to check the resistance of a resistor which is marked 1 KΩ. The circuit is connected as in Fig. 213, if the voltmeter has a resistance of 1 KΩ find the resistance value from the calculated instrument readings.

Fig. 213 Ammeter/voltmeter method for measuring resistance.

The resistance of the ammeter may be assumed to be negligible. State the error in the calculated value.

(Ans. 51.55 V, 6.48 V).

12. A moving coil instrument gives full-scale deflection when the current through it is 10 mA and the p.d. across it is 5 V.

   (i) Calculate the resistance of the instrument.

   (ii) Find the value of resistor required to enable the instrument to read 38 V at full-scale deflection and illustrate your solution with a diagram.

   (iii) Find the value of resistor required to enable the instrument to read 10 A at full scale deflection and illustrate your solution with a diagram.

   (Ans. (i) 500 Ω (ii) 3.3 KΩ (iii) 0.505 Ω).

13. Construct a simple ohmmeter using an ammeter with a full-scale deflection of 15 mA a variable resistor, and a 15 V battery. Draw the circuit for the arrangement.

   Calculate the resistance indicated by the instrument when currents of (i) 1 mA (ii) 5 mA and (iii) 10 mA are flowing in the circuits.

   (Ans. (i) 14 KΩ (ii) 2 KΩ (iii) 0.5 KΩ).

14. Suggest a suitable method of measuring each of the following:

   (i) The r.m.s. value of a 100 Hz sinusoidal voltage of the order of 10 V.

   (ii) The r.m.s. value of a 250 Hz sinusoidal voltage of the order of 100 mV.

   (iii) The peak value of a voltage pulse of rectangular waveform of duration about 20 μs and repeating every 100 μs.

PART II

SOLUTIONS 1

1. The total resistance is calculated from the formula

$$\frac{1}{R_T} = \frac{1}{R_1} + \frac{1}{R_2} + \frac{1}{R_3}$$

$$\frac{1}{R_T} = \frac{1}{1} + \frac{1}{2} + \frac{1}{3} = \frac{6 + 3 + 2}{6} = \frac{11}{6}$$

$$R_T = \frac{6}{11}\ \Omega.$$

Fig. 214 shows the circuit.

Fig. 214 Parallel resistors.

The p.d. between A and B, $V_{AB} = 12 \times \dfrac{6}{11} = 6.55$ V. The currents I_1, I_2

and I_3 can be determined $I_1 = \dfrac{V_{AB}}{1} = 6.55$ A, $I_2 = \dfrac{V_{AB}}{2} = 3.27$ A,

$$I_3 = \frac{V_{AB}}{3} = 2.18\ \text{A}.$$

2. The circuit is shown in Fig. 215.

Fig. 215 Current divider

166

Using the current divider formulae

$$I_1 = \frac{R_2}{R_1 + R_2}I = \frac{5}{8} \times 1 = \frac{5}{8} = 0.625 \text{ A.}$$

$$I_2 = \frac{R_1}{R_1 + R_2}I = \frac{3}{8} \times 1 = \frac{3}{8} = 0.375 \text{ A.}$$

3.   Fig. 216 shows the circuit

Fig. 216   Current divider.

$$I_2 = \frac{R_1}{R_1 + R_2}I = \frac{2}{10} \times 5 = 1 \text{ A} \qquad I_1 = \frac{R_2}{R_1 + R_2}I = \frac{8}{10} \times 5 = 4 \text{ A.}$$

4.   Fig. 217 shows the circuit.

Fig. 217   Current divider.

$$I_1 = \frac{10}{10 + 20} \times 3.6 = \frac{1}{3} \times 3.6 = 1.2 \text{ A}$$

$$I_2 = \frac{20}{10 + 20} \times 3.6 = \frac{2}{3} \times 3.6 = 2.4 \text{ A.}$$

167

5. $\dfrac{1}{R_T} = \dfrac{1}{30} + \dfrac{1}{40} + \dfrac{1}{50} = \dfrac{20 + 15 + 12}{600} = \dfrac{47}{600}$

$R_T = \dfrac{600}{47} = 12.8 \ \Omega.$

The p.d. across the input $= 12.8 \times 2 = 25.6$ V

$I_1 = \dfrac{25.6}{30} = 0.85$ A, $I_2 = \dfrac{25.6}{40} = 0.64$ A, $I_3 = \dfrac{25.6}{50} = 0.51$ A.

6. $\dfrac{1}{R_T} = \dfrac{1}{3} + \dfrac{1}{7} + \dfrac{1}{R} = \dfrac{1}{1}$

$\dfrac{1}{R} = 1 - \dfrac{1}{3} - \dfrac{1}{7} = 0.5238$

$R = \dfrac{1}{0.5238} = 1.91 \ \Omega.$

$V_T =$ p.d. across $R = 10 \times 1.91 = 19.1$ V

$I_1 = \dfrac{19.1}{3} = 6.37$ A $\qquad I_2 = \dfrac{19.1}{7} = 2.73$ A

$\dfrac{V}{R_T} = I = 6.37 + 2.73 + 10 = 19.1$ A.

7. The total current $= I = 0.1 + 0.2 + 0.3 + 0.4 = 1$ A.
   The total voltage $= 1 \times 10 = 10$ V

   $R_1 = \dfrac{10}{0.1} = 100 \ \Omega, \qquad R_2 = \dfrac{10}{0.2} = 50 \ \Omega.$

   $R_3 = \dfrac{10}{0.3} = 33.3 \ \Omega, \qquad R_4 = \dfrac{10}{0.4} = 25 \ \Omega.$

8. Fig. 218 shows the circuit.

Fig. 218 Shunt.

168

$$5I = (15 - I)0.0005$$

$$5I + 0.0005I = 0.0075$$

$$I = \frac{0.0075}{5.0005} = 1.49985 \times 10^{-3}$$

$$I = 1.5 \text{ mA approximately.}$$

9. Fig. 219 shows the circuit

Fig. 219 Loaded potential divider.

$$\frac{\frac{R}{2} \times 1500}{\frac{R}{2} + 1500} I = 35 \qquad \dots (1)$$

$$I \left(\frac{R}{2} + \frac{\frac{R}{2} \times 1500}{\frac{R}{2} + 1500} \right) = 100 \quad \dots (2)$$

Dividing equation (2) by equation (1) we have

$$\frac{100}{35} = \frac{\frac{R}{2} + \frac{\frac{R}{2} \times 1500}{\frac{R}{2} + 1500}}{\frac{\frac{R}{2} \times 1500}{\frac{R}{2} + 1500}}$$

$$= \frac{\frac{R}{2} + 1500}{1500} + 1$$

169

$$\frac{100}{35} = \frac{R}{2 \times 1500} + 1 + 1$$

$$\frac{100}{35} - 2 = \frac{R}{3000}$$

$R = 2571.4 \ \Omega$ hence $R \approx 2570 \ \Omega$.

10. $\dfrac{R_1 10000}{R_1 + 10000} I_1 = 80 \qquad \ldots (1)$

$R_2 I_1 = 120 \qquad \ldots (2)$

Dividing (1) by (2)

$$\frac{R_1 10000}{(R_1 + 10000) R_2} = \frac{2}{3} \qquad \ldots (3)$$

$$I_2 \times \frac{R_2 \times 10000}{R_2 + 10000} = 100 \qquad \ldots (4)$$

$$I_2 R_1 = 100 \qquad \ldots (5)$$

Dividing (4) by (5)

$$\frac{R_2 10000}{R_1 (R_2 + 10000)} = 1 \qquad \ldots (6)$$

From (3) $\quad 30000 R_1 = 2 R_1 R_2 + 20000 R_2 \ \ldots (7)$

From (6) $\quad 10000 R_2 = R_1 R_2 + 10000 R_1 \quad \ldots (8)$

Substituting $2 \times (8)$ in (7)

$\quad 30000 R_1 = 2 R_1 R_2 + 2 R_1 R_2 + 20000 R_1$

$10000 R_1 = 4 R_1 R_2$

$R_2 = 2500 \ \Omega$

and substituting this value in (8)

$10000 \times 2500 \ = \ 2500 R_1 + 10000 R_1$

$10000 \times 2500 \ = \ 12500 R_1$

$$R_1 = \frac{10000 \times 2500}{12500} = 2000 \ \Omega.$$

11. $V = \dfrac{5}{30} \times 10 = \dfrac{5}{3} = 1.67$ V.

12. $V = \dfrac{5000}{30000} \times 100 = \dfrac{50}{3} = 16.7$ V.

13. Fig. 220 shows the circuit.

Fig. 220 Ladder network.

(i) $\dfrac{20 \times 10}{30} = \dfrac{20}{3}$ $\qquad V_1 = \dfrac{20/3}{5 + 20/3} \times 10 = \dfrac{20/3}{35/3} \times 10 = \dfrac{200}{35}$

$V = \dfrac{15}{20} \times \dfrac{200}{35} = \dfrac{150}{35} = 4.29$ V.

(ii) The total resistance of the circuit, $R_T = 5 + \dfrac{20}{3} = \dfrac{35}{3}$ Ω, the total

current, $I = \dfrac{10}{35/3} = \dfrac{6}{7}$ A.

The current through the 5 Ω and 15 Ω resistors in series is

$\dfrac{10}{30} \times \dfrac{6}{7} = \dfrac{2}{7}$ A, and $V = 15 \times \dfrac{2}{7} = \dfrac{30}{7} = 4.29$ volts.

14.

I(A)	6	4.8	4	3.43	3	2.67	2.4	2.18	2
R(Ω)	0	$\frac{1}{4}$	$\frac{1}{2}$	$\frac{3}{4}$	1	$1\frac{1}{4}$	$1\frac{1}{2}$	$1\frac{3}{4}$	2
P(W)	0	5.76	8	8.82	9	8.89	8.64	8.33	8

For $R = \dfrac{1}{4}$ Ω, $I = \dfrac{E}{r + R} = \dfrac{6}{1 + 1/4} = 4.8$ A

$$P = I^2R = 4.8^2 \left(\dfrac{1}{4}\right) = 5.76 \text{ W}$$

171

For $R = \dfrac{3}{4}$ Ω, $I = \dfrac{E}{r + R} = \dfrac{6}{1 + 3/4} = 3.43$ A

$$P = I^2R = 3.43^2 \left(\dfrac{3}{4} \right) = 8.82 \text{ W}$$

For $R = 1\dfrac{1}{4}$ Ω, $I = \dfrac{E}{r + R} = \dfrac{6}{1 + 1\dfrac{1}{4}} = 2.67$ A

$$P = I^2R = 2.67^2 \left(1\dfrac{1}{4} \right) = 8.89 \text{ W}$$

For $R = 1\dfrac{3}{4}$ Ω, $I = \dfrac{E}{r + R} = \dfrac{6}{1 + 1\dfrac{3}{4}} = 2.18$ A

$$P = I^2R = 2.18^2 \times 1\dfrac{3}{4} = 8.33 \text{ W.}$$

Fig. 221 shows the graph of P against R.

Fig. 221 Graph of power against load resistance.
Maximum Power Transfer.

The maximum power dissipation in the load is 9 W when $R = 1$ Ω.

Therefore the maximum power dissipation occurs in the load when the load resistance is equal to the internal resistance.

172

15. Referring to Fig. 41, the p.d. across the 6 Ω resistor = 6 V

the current through the 6 Ω resistor = $\dfrac{6}{6}$ = 1 A

I_1 = 1 A

the total current = $I = I_1 + 0.5 = 1 + 0.5 = 1.5$ A

the p.d. across $R = 15 - 6 = 9$ V therefore

$$R = \frac{9}{1.5} = 6\ \Omega.$$

16. (a) $R_{AB} = R_1 + \dfrac{R_2 R_3}{R_2 + R_3} + \dfrac{R_4 R_5 R_6}{R_4 R_5 + R_4 R_6 + R_5 R_6}$

$= 1 + \dfrac{1 \times 1}{1 + 1} + \dfrac{3 \times 3 \times 3}{3 \times 3 + 3 \times 3 + 3 \times 3}$

$= 1 + \dfrac{1}{2} + 1 = 2.5\ \Omega.$

(b) (i) $I_1 = \dfrac{10}{R_{AB}} = \dfrac{10}{2.5} = 4$ A

(ii) $I_2 = \dfrac{I_1}{2} = 2$ A

(iii) $I_3 = \dfrac{I_1}{3} = \dfrac{4}{3} = 1.33$ A.

SOLUTIONS 2

1. Redrawing Fig. 71, we have Fig. 222 showing the currents and p.d.s, through and across respectively.

Fig. 222 Kirchhoff's Laws.

Assume that the currents flowing in the 5 V and 3 V batteries are I_1 and I_2 respectively, and hence the load current through the 5 Ω load is $I_1 + I_2$ by applying KCL. Mark the p.d.s by arrows as shown. Applying KVL to the loops *ABCDEFA* and *BCDEB*, we have

$$5 - I_1 - \left(I_1 + I_2\right)5 = 0 \quad \text{or} \quad 6I_1 + 5I_2 = 5 \qquad \ldots (1)$$

$$3 - 2I_2 - \left(I_1 + I_2\right)5 = 0 \quad \text{or} \quad -5I_1 - 7I_2 = -3 \quad \ldots (2)$$

Multiplying equation (1) by 5 and equation (2) by 6, we have

$$30I_1 + 25I_2 = 25 \qquad \ldots (3)$$

$$-30I_1 - 42I_2 = -18 \quad \ldots (4)$$

Adding equations (3) and (4)

$$-17I_2 = 7$$

$$I_2 = -\frac{7}{17} = -0.41 \text{ A}$$

the negative sign indicates that the assumed current I_2 is flowing in the opposite direction. Substituting $I_2 = -0.41$ A in equation (1)

$$6I_1 + 5I_2 = 5$$

$$6I_1 - 5 \times 0.41 = 5$$

174

$6I_1 = 7.05$

$I_1 = 1.175$ A

the assumed direction of I_1 is correct, therefore I_1 is the discharging current.

$I_1 + I_2 = 1.175 - 0.41 = 0.765$ A.

Therefore $I_1 = 1.175$ A discharging current

$I_2 = -0.41$ A charging current

$I_1 + I_2 = 0.765$ A the load current.

2. Fig. 223 is the redrawn circuit showing the currents and voltages

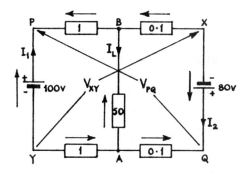

Fig. 223 Kirchhoff's Laws.

(a) Let I_1 and I_2 be the direction of the currents in the batteries and I_L is the load current. At A, $I_L + I_2 = I_1$, therefore $I_L = I_1 - I_2$.

Applying KVL to the loops *PBAYP* and *BXQAB* we have for the loop *PBAYP*

$$100 = I_1 + (I_1 - I_2)50 + I_1 \text{ or}$$

$$100 = 52I_1 - 50I_2$$

simplifying $50 = 26I_1 - 25I_2$ ... (1)

for the loop *BXQAB*

$80 + (I_1 - I_2)50 = 0.1I_2 + 0.1I_2$

$80 = -50I_1 + 50.2I_2$ or

$40 = -25I_1 + 25.1I_2$ ... (2)

175

Multiplying equation (1) by 25 and equation (2) by 26 we have

$1250 = 650I_1 - 625I_2$ ... (3)

$1040 = -650I_1 + 652.6I_2$ ... (4)

adding (3) and (4)

$2290 = 27.6I_2$

$I_2 = 83.0$ A

substituting this value in (1)

$50 = 26I_1 - 25 \times 83.0$

$26I_1 = 50 + 2075$

$I_1 = \dfrac{2125}{26} = 81.7$ A

$I_L = I_1 - I_2 = 81.7 - 83.0 = -1.30$ A.

The assumed direction of the load current is incorrect − 1.30 A flows in the opposite direction indicated, through the 50 Ω.

Therefore $I_1 = 81.7$ A, $I_2 = 83.0$ A, $I_L = -1.30$ A.

(b)   Let V_{PQ} be the p.d. between P and Q and let V_{XY} be the p.d. between X and Y. Applying KVL to the loops $PQAYP$ and $XQAYX$, we have

$$1I_1 + 0.1I_2 + V_{PQ} = 100$$

therefore $V_{PQ} = 100 - I_1 - 0.1I_2$

$= 100 - 81.7 - 0.1 \times 83.0$

$= 100 - 81.7 - 8.3 = 100 - 90$

$= 10$ volts

$V_{XY} + 80 = 0.1I_2 + 1I_1$

$V_{XY} = 0.1I_2 + I_1 - 80 = 0.1 (83) + 81.7 - 80$

$= 8.3 + 81.7 - 80 = 8.3 + 1.7$

$= 10$ volts.

8. Fig. 73 is now redrawn and it is equivalent to Fig. 224.

Fig. 224 Kirchhoffs Laws. Equivalent circuit.

Applying KVL to the loops *DACBED* and *CGFBC* we have

DACBED $200 = 25I_1 + 75(I_1 + I_2)$

CGFBC $50 = 2I_2 + (I_1 + I_2)75$

simplifying $100I_1 + 75I_2 = 200$

dividing each term by 25

$\qquad 4I_1 + 3I_2 = 8 \qquad\qquad$ ... (1)

$\qquad 75I_1 + 77I_2 = 50 \qquad\qquad$ ... (2)

Multiplying (1) by 75 and (2) by − 4

$\qquad 300I_1 + 225I_2 = 600 \qquad$ ... (3)

$\qquad - 300I_1 - 308I_2 = - 200 \qquad$ ... (4)

Adding (3) and (4)

$\qquad - 83I_2 = 400$

$\qquad I_2 = - 4.82 \text{ A}$

substituting this value in (1)

$\qquad 4I_1 + 3(- 4.82) = 8$

$\qquad 4I_1 = 8 + 3 (4.82)$

$\qquad 4I_1 = 22.5$

$\qquad I_1 = 5.63 \text{ A}.$

4.   Fig. 225 shows the direction of currents and the p.d.s.

Fig. 225 Kirchhoff's Laws.

Applying KCL at the junction A

$I_1 + I_2 = I_L$

Applying KVL to loop 1

$2 = 0.1I_1 + I_L$

or $0.1I_1 + I_1 + I_2 = 2$

or $1.1I_1 + I_2 = 2$ ... (1)

Applying KVL to loop 2

$3 = 0.2I_2 + I_L$

or $0.2I_2 + I_1 + I_2 = 3$

$I_1 + 1.2I_2 = 3$ ... (2)

Multiplying equation (1) by 1 and (2) by (− 1.1)

$1.1I_1 + I_2 = 2$

$- 1.1I_1 - 1.2(1.1)I_2 = 3(- 1.1)$

or $1.1I_1 + I_2 = 2$ ... (3)

$- 1.1I_1 - 1.32I_2 = - 3.3$ ... (4)

Adding (3) and (4)

$- 0.32I_2 = - 1.3$

$I_2 = 4.06$ A

substituting $I_2 = 4.06$ in (2)

$I_1 = 3 - 1.2 (4.06)$

$I_1 = - 1.87$ A

I_1 flows in the opposite direction indicated.

178

5.   Fig. 226 shows the circuit with the currents and the voltages.

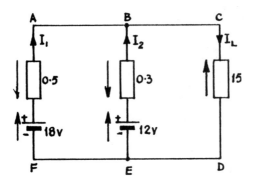

Fig. 226

At the junction B $I_L = I_1 + I_2$

Applying KVL to the loops $ABEFA$ and $BCDEB$, we have

$$18 - 12 = 0.5I_1 - 0.3I_2$$
or $0.5I_1 - 0.3I_2 = 6$ ... (1)

$BCDEB$ $12 = 0.3I_2 + 15(I_1 + I_2)$
    or $15I_1 + 15.3I_2 = 12$ ... (2)

Multiplying equation (1) by -15 and equation (2) by 0.5 we have

$$-7.5I_1 + 4.5I_2 = -90$$ ... (3)
$$7.5I_1 + 7.65I_2 = 6$$ ... (4)

Adding equations (3) and (4)

$$12.15I_2 = -84$$
$$I_2 = -6.91 \text{ A}$$

substituting this value in (1)

$$0.5I_1 - 0.3(-6.91) = 6$$
$$0.5I_1 = 6 - 2.07$$

$$I_1 = \frac{3.93}{0.5} = 7.86 \text{ A}$$

$I_L = I_1 + I_2 = 7.86 - 6.91 = 0.95$ A.

Therefore $I_1 = 7.86$ A, $I_2 = -6.91$ A and $I_L = 0.95$ A.

179

6.   Referring to Fig. 50, $I = \dfrac{40 + 25 - 10 - 20}{1 + 2 + 3} = 5.83$ A.

7.   Referring to Fig. 52, $I_1 + I_2 + I_3 = 0$ hence $I_3 = -I_1 - I_2$.

Loop *ADBEA* $3 - 2 + 2I_1 - 5I_2 = 0$ or $2I_1 - 5I_2 = -1$ ... (1)

Loop *ADBCA* $-3 + 5I_2 - 10I_3 + 5 = 0$

or $5I_2 - 10(-I_1 - I_2) = -2$

or $10I_1 + 15I_2 = -2$ ... (2)

Multiplying equation (1) by 3, we have

$6I_1 - 15I_2 = -3$ ... (3)

$10I_1 + 15I_2 = -2$ ... (2)

adding (3) and (2)

$16I_1 = -5$

$I_1 = -\dfrac{5}{16} = -0.3125$ A.

Substituting this value in (1)

$2(-0.3125) - 5I_2 = -1$

$5I_2 = -0.625 + 1 = 0.375$

$I_2 = 0.075$ A

$I_3 = -(I_1 + I_2) = -(-0.3125 + 0.075) = 0.2375$ A.

The actual currents flow as shown in the diagram of Fig. 227.

Fig. 227 Actual currents Kirchhoff's Laws.

0.075 A and 0.2375 A are the discharging currents and 0.3125 A is the charging current.

180

8.   Referring to Fig. 53 and assuming that the current flow from the positive terminal of the batteries as shown by I_1 and I_2, then the current through 100 Ω is $I_1 - I_2$ since $I_2 + \left(I_1 - I_2\right) = I_1$ at the junction A

Loop *ABDEA* $\qquad 10 + 100\left(I_1 - I_2\right) = 0$

$$\text{or } 100I_1 - 100I_2 = -10$$

$$\text{or } 10I_2 - 10I_1 = 1 \quad \dots \text{ (1)}$$

Loop *ACBA* $\qquad -30 + 10I_1 = 10$

$$40 = 10I_1$$

$$I_1 = 4 \text{ A}$$

substituting in (1)

$$10I_2 - 4\left(10\right) = 1$$

$$10I_2 = 41$$

$$I_2 = 4.1 \text{ A}$$

$I_1 - I_2 = 4 - 4.1 = -0.1 \text{ A}$

the current should flow in the opposite direction from that indicated in the diagram.

9.   Referring to Fig. 54,

loop *ACBA* $\qquad 2.5 - 1.5I_1 + \left(I_1 - I_2\right)15 = 0$

$$\text{or } 13.5I_1 - 15I_2 = -2.5$$

$$\text{or } 15I_2 - 13.5I_1 = 2.5 \qquad \dots \text{ (1)}$$

Loop *ABDA* $\qquad -15\left(I_1 - I_2\right) + 3.5 - 1.5I_2 = 0$

$$-15I_1 + 15I_2 + 3.5 - 1.5I_2 = 0$$

$$-15I_1 + 13.5I_2 = -3.5$$

$$\text{or } 15I_1 - 13.5I_2 = 3.5 \qquad \dots \text{ (2)}$$

Multiplying (1) by 15 and (2) by 13.5, we have

$$-202.5I_1 + 225I_2 = 37.5 \qquad \dots \text{ (3)}$$

$$202.5I_1 - 182.25I_2 = 47.25 \qquad \dots \text{ (4)}$$

181

adding (3) and (4)

$$42.75I_2 = 84.75$$

$$I_2 = 1.98 \text{ A.}$$

Substituting this value in (2)

$$15I_1 - 13.5 \ (1.98) = 3.5$$

$$15I_1 = 3.5 + 26.7$$

$$I_1 = 2.02 \text{ A}$$

$$I_1 - I_2 = 2.02 - 1.98 = 0.04 \text{ A}$$

The p.d. across A and B, $V_{AB} = 15 \times 0.04 = 0.60$.

10. Referring to Fig. 57, we have

$$I_1 + I_2 + I_3 = 0$$

$$I_3 = -\left(I_1 + I_2\right)$$

Applying KVL

Loop *EBDAE* $50 - 20 + 5I_2 - I_1 = 0$

or $I_1 - 5I_2 = 30$ ... (1)

Loop *CBDAC* $100 - 20 + 5I_2 + \left(I_1 + I_2\right)0.1 = 0$

$$0.1I_1 + 5.1I_2 = -80 \quad ... (2)$$

Multiplying (2) by 10

$$I_1 + 51I_2 = -800 \quad ... (3)$$

multiplying (1) by (– 1)

$$-I_1 + 5I_2 = -30 \quad ... (4)$$

Adding (3) and (4)

$$56I_2 = -830$$

$$I_2 = -14.8 \text{ A}$$

substituting in (1)

$$I_1 = 30 + 5 \ (-14.8) = 30 - 74 = -44 \text{ A}$$

$$I_3 = -\left(I_1 + I_2\right) = -(-44 - 14.8) = 58.8 \text{ A.}$$

The currents are $I_1 = -44$ A, $I_2 = -14.8$ A, $I_3 = 58.8$ A the actual currents flow are as shown in Fig. 228.

Fig. 228 Kirchhoff's Laws. Actual current.

1. Referring to Fig. 58

$$R_T = 1 + \frac{5 \times 15}{5 + 15} = 1 + \frac{75}{20} = 4.75 \ \Omega$$

$$I = \frac{25}{4.75} = 5.26 \text{ A}$$

using the current divider

$$I_1 = \frac{15}{5 + 15} \times 5.26 = 3.95 \text{ A}$$

$$I_2 = \frac{5}{20} \times 5.26 = 1.32 \text{ A or}$$

$$I_2 = I - I_1 = 5.26 - 3.945 = 1.315 \text{ A}$$

$$V_{AB} = 5 \times I_1 = 15 \times I_2 = 5 \times 3.945 = 19.73 \text{ volts}$$

2. Referring to Fig. 59

The total resistance of the load $= \dfrac{10 \times 20}{10 + 20} = \dfrac{20}{3} \ \Omega.$

The circuit of Fig. 59 is now equivalent to the circuit of Fig. 229.

Fig. 229 Kirchhoffs Laws.

Loop *CABC* $200 = I_1 + \dfrac{20}{3}\left(I_1 + I_2\right)$

$$\frac{23}{3}I_1 + \frac{20}{3}I_2 = 200$$

$$23I_1 + 20I_2 = 600 \quad \ldots (1)$$

Loop *BDAB* $100 = 2I_2 + \dfrac{20}{3}\left(I_1 + I_2\right)$

or $\dfrac{20}{3}I_1 + \dfrac{26}{3}I_2 = 100$

or $20I_1 + 26I_2 = 300 \quad \ldots (2)$

Multiplying (1) by 20 and (2) by -23 we have

$$460I_1 + 400I_2 = 12000 \quad \ldots (3)$$
$$-460I_1 - 598I_2 = -6900 \quad \ldots (4)$$

adding (3) and (4)

$$-198I_2 = 5100$$
$$I_2 = -25.8 \text{ A}$$

substituting in (1)

$$23I_1 + 20\,(-25.8) = 600$$
$$23I_1 = 600 + 516$$
$$I_1 = 48.5 \text{ A}$$

184

13. Referring to Fig. 61, we have

$$I = \frac{25 - 15}{7} = 1.43 \text{ A}$$

$$V_{AB} = 25 - 2I = 25 - 2 \times 1.43$$

$$V_{AB} = 25 - 2.86 = 22.14 \text{ volts.}$$

14. Referring to Fig. 60, we have. Let $3I$ be the supply current, since the circuit is symmetrical, I flows in AB, I flows in AD and I flows in AE, then $\frac{I}{2}$ flows in BC and BF, $\frac{I}{2}$ flows in DC and DH, $\frac{I}{2}$ flows in EF and EH and the total current flows out at G as $3I$.

Consider the loop $ABCGA$

$$V = I \times 1 + \frac{I}{2} \times 1 + I \times 1$$

$$R = \frac{V}{3I} = \frac{1}{3} + \frac{1}{6} + \frac{1}{3} = \frac{2}{3} + \frac{1}{6} = \frac{5}{6} \ \Omega$$

$$R = \frac{5}{6} \ \Omega.$$

15. Fig. 230 shows the circuit required

Fig. 230 Kirchhoff's Laws.

Applying KVL to loop $ABEFCDA$

$$5 = I_1 + (I_1 + I_2)4.5$$
$$5.5I_1 + 4.5I_2 = 5 \quad \dots (1)$$

185

Applying KVL to loop *BEFCB*

$$10 = 5.5I_2 + 4.5(I_1 + I_2)$$
$$4.5I_1 + 10I_2 = 10 \quad \dots (2)$$

Multiplying (1) by 4.5 and (2) by -5.5, we have

$$24.75I_1 + 20.25I_2 = 22.5 \quad \dots (3)$$
$$-24.75I_1 - 55I_2 = -55 \quad \dots (4)$$

adding (3) and (4)

$$-34.75I_2 = -32.5$$
$$I_2 = 0.935 \text{ A}$$

substituting this value in (1)

$$5.5I_1 + 4.5 (0.935) = 5$$
$$5.5I_1 = 5 - 4.2075$$
$$I_1 = 0.144 \text{ A}$$
$$I_1 + I_2 = 0.144 + 0.935 = 1.079 \text{ A}.$$

Therefore the branch currents are:

$$I_1 = 0.144 \text{ A}, I_2 = 0.935 \text{ A}, I_1 + I_2 = 1.079 \text{ A}.$$

16. Fig. 231 shows the circuit required

Fig. 231 Kirchhoff's Laws.

KVL ABCDA loop $110 = 20I_1 + 100I_1 + 100I_2$

or $\quad 12I_1 + 10I_2 = 11 \quad \dots (1)$

Loop *BEFCB* loop

$$120 = 40I_2 + 100I_1 + 100I_2$$
$$100I_1 + 140I_2 = 120$$

or $\quad 5I_1 + 7I_2 = 6 \quad \dots (2)$

186

Multiplying (1) by 5 and (2) by (– 12), we have

$$60I_1 + 50I_2 = 55 \quad \ldots (3)$$

$$- 60I_1 - 84I_2 = - 72 \quad \ldots (4)$$

adding (3) and (4)

$$- 34I_2 = - 17$$

$$I_2 = \frac{17}{34} = \frac{1}{2}$$

$$I_2 = 0.5 \text{ A}$$

substituting value in (1)

$$12I_1 + 10 \ (0.5) \ = \ 11$$

$$12I_1 \ = \ 6$$

$$I_1 \ = \ 0.5 \text{ A}$$

$$I_1 + I_2 = 0.5 + 0.5 = 1 \text{ A}.$$

The p.d. across the 100 Ω resistor load = $1 \times 100 = 100$ V.

7. Fig. 232 shows the circuit.

Fig. 232 Kirchhoff's Law.

Applying KVL to the loops *ABEFA* and *BCDEB*

Loop *ABEFA* $10 - 5 + 1.5I_2 - I_1 = 0$

or $I_1 - 1.5I_2 = 5 \qquad \ldots (1)$

Loop *BCDEB* $5 = 1.5I_2 + 10I_1 + 10I_2$

or $10I_1 + 11.5I_2 = 5 \ldots (2)$

187

Multiplying (1) by – 10 and (2) by 1, we have

$$- 10I_1 + 15I_2 = - 50 \quad \dots (3)$$
$$10I_1 + 11.5I_2 = 5 \quad \dots (4)$$

adding (3) and (4)

$$26.5I_2 = - 45$$

$$I_2 = - \frac{45}{26.5} = - 1.7 \text{ A}$$

substituting this value to (1)

$$I_1 - 1.5 (- 1.7) = 5$$

$$I_1 = 5 - 2.55$$

$$I_1 = 2.45 \text{ A}$$

$$I_1 + I_2 = 2.45 - 1.7 = 0.75 \text{ A}$$

$$I_L = 0.75 \text{ A}$$

SOLUTIONS 3

1. $C_1 = \dfrac{\in_o \in_r A_1}{d_1} \quad \dots (1) \qquad C_2 = \dfrac{\in_o \in_r A_2}{d_2} \quad \dots (2)$

the formulae for the capacitances are given. Dividing (1) by (2)

$$\frac{C_1}{C_2} = \frac{A_1}{A_2} \frac{d_2}{d_1}.$$

If $C_1 = 100$ pF, $C_2 = C_1 \dfrac{A_2 d_1}{A_1 d_2} = \dfrac{C_1 d_1 A_1 /2}{A_1 2d_1} = \dfrac{1}{4} C_1$

$$C_2 = \frac{1}{4} 100 \text{ pF} = 25 \text{ pF}$$

$$C_2 = 25 \text{ pF}.$$

2.   (a)   The capacitance of a capacitor depends directly on the area of the plates, A, directly on the dielectric constant of the material placed between the plates, $\in_r$, and is inversely proportional to the distance of separation, d.

   (b) $C = \dfrac{\in_o \in_r A}{d}$

188

C = capacitance in farads

ϵ_o = 8.85 × 10⁻¹² F/m permittivity of free space

ϵ_r = relative permittivity (no units)

A = area of the plates (m^2)

d = separation (m)

(c) (i) $Q = CV$ Q (coulombs), C (farads), V (volts)

(ii) $W = \dfrac{1}{2}CV^2$ W (joules), C (farads), V (volts).

$$C = \frac{\epsilon_o A}{d}$$

$$5 \times 10^{-12} = \frac{\epsilon_o A}{2.5 \times 10^{-3}} \quad \dots (1)$$

$$C = \frac{\epsilon_o A}{3.5 \times 10^{-3}} \quad \dots (2)$$

Dividing (2) by (1)

$$\frac{C}{5 \times 10^{-12}} = \frac{2.5 \times 10^{-3}}{3.5 \times 10^{-3}}$$

$$C = \frac{2.5}{3.5} \times 5 \times 10^{-12} = 3.57 \times 10^{-12} \text{ F}$$

$$C = 3.57 \text{ pF}$$

$$C_1 = \frac{\epsilon_o \epsilon_{r_1} A}{d} \quad \dots (1) \qquad\qquad C_2 = \frac{\epsilon_o \epsilon_{r_2} A}{d} \quad \dots (2)$$

A and d are unaffected, dividing (1) by (2) we have $\dfrac{C_1}{C_2} = \dfrac{\epsilon_{r_1}}{\epsilon_{r_2}}$

$$C_2 = C_1 \frac{\epsilon_{r_2}}{\epsilon_{r_1}} = 5 \times 10^{-12} \times \frac{5}{1} = 25 \times 10^{-12} \text{ F}$$

$$C_2 = 25 \text{ pF}.$$

Increasing the permittivity, increases the value of the capacitance.

4. $d = 2.5$ mm $= 2.5 \times 10^{-3}$ m $\qquad A = 100$ cm$^2 = 100 \times 10^{-4}$ m^2

$C = 100$ pF

(i) electric flux $= Q = CV = 100 \times 10^{-12} \times 15000$

$\qquad\qquad\qquad\qquad = 15 \times 10^{-7} = 1.5 \ \mu C$

(ii) electric flux density $= D = \dfrac{Q}{A} = \dfrac{1.5 \times 10^{-6}}{100 \times 10^{-4}} = 1.5 \times 10^{-4}$ C/m^2

(iii) $\in = \dfrac{D}{E} = \dfrac{1.5 \times 10^{-4}}{15000/2.5 \times 10^{-3}} = \dfrac{1.5 \times 10^{-4}}{6 \times 10^{6}} = 25 \times 10^{-12}$ F/m

$\in \ = \ \in_r \in_o \ = 25 \times 10^{-12} = \ \in_r 8.85 \times 10^{-12}$

$\in_r \ = 2.83.$

5. $W = \dfrac{1}{2} QV = 10^{-6}$

$QV = 2 \times 10^{-6}$

$V = \dfrac{2 \times 10^{-6}}{100 \times 10^{-6}}$

$V = 0.02$ volts

$C = \dfrac{Q}{V} = \dfrac{100 \times 10^{-6}}{0.02}$

$C = 5000 \ \mu F.$

6. Fig. 233 shows the circuit of two capacitors in series.

Fig. 233 Two capacitors in series.

(i) The total charge, Q, is found $Q = CV = 6.67 \times 10^{-6} \times 20$

$Q = 133.3\ \mu C$ where $C = \dfrac{10 \times 20}{10 + 20}\ \mu F = 6.67\ \mu F$.

The p.d. across $A = \dfrac{Q}{C_A} = \dfrac{133.3 \times 10^{-6}}{10 \times 10^{-6}} = 13.3\ V$

The energy stored in $A = \dfrac{1}{2} C_A V_A^2 = \dfrac{1}{2} \times 10 \times 10^{-6} \times 13.33^2$

$$= 888\ \mu J.$$

(ii) Fig. 234 shows the circuit of two capacitors in parallel.

Fig. 234 Two capacitors in parallel.

The p.d. across $C_A = 20\ V$.

The energy stored in $A = \dfrac{1}{2} C_A V_A^2 = \dfrac{1}{2} \times 10 \times 10^{-6} \times 20^2$

$$= 2\ mJ.$$

7. (i) The equivalent capacitance of the circuit of Fig. 235 is given by

Fig. 235 series-parallel capacitors.

191

$$C = \left[\frac{10 \times 30}{10 + 30} + 20 \right] \, \mu F$$

$$C = 27.5 \, \mu F.$$

(ii)   The charge on the 30 μF capacitor is found

$$Q = \left[\frac{10 \times 30}{10 + 30} \right] 100 \times 10^{-6} = 750 \, \mu C$$

the p.d. across the 30 $\mu F = \dfrac{Q}{C} = \dfrac{750 \times 10^{-6}}{30 \times 10^{-6}} = 25 \text{ V.}$

(iii) The charge on the 10 μF capacitor is the same as that across the 30 μF, 750 μC.

(iv) $W = \dfrac{1}{2} C V^2 = \dfrac{1}{2} \times 20 \times 10^{-6} \times 100^2 = 0.1 \text{ J.}$

8.   (i)   Fig. 236 is the circuit required with the appropriate switches.

Fig. 236 Charge sharing in capacitors.

S_1 switch is closed and the 1 μF capacitor is charged up to 100 V, S_1 is opened and S_2 is closed what is the p.d. across the two capacitors?

$Q = CV$

$Q = 1 \times 10^{-6} \times 100 = 100 \, \mu C$ is the charge on C_1.

Let V_o be the new value of the p.d. required.

$$V_o = \frac{Q}{C_1 + C_2} = \frac{100 \times 10^{-6}}{3 \times 10^{-6}} = 33.3 \text{ V.}$$

192

(ii) The energy stored in each capacitor before paralleling

$$W_1 = \frac{1}{2} C_1 V^2 = \frac{1}{2} \times 1 \times 10^{-6} \times 100^2 = 0.005 \text{ J}$$

$$W_2 = \frac{1}{2} C_2 0^2 = 0 \text{ J}$$

the energy stored in each capacitor after paralleling.

$$W_1 = \frac{1}{2} C_1 V_o^2 = \frac{1}{2} \times 1 \times 10^{-6} \times 33.3^2 = 554 \ \mu\text{J}$$

$$W_2 = \frac{1}{2} C_2 V_o^2 = \frac{1}{2} \times 2 \times 10^{-6} \times 33.3^2 = 1109 \ \mu\text{J}.$$

Total energy before paralleling 5 mJ.

Total energy after paralleling 1.66 mJ.

A loss of energy $(5 - 1.66)$ mJ $= 3.34$ mJ due to heat energy in a form of spark.

(iii) Fig. 237 shows the two capacitors in series.

Fig. 237 Capacitors in series.

$$Q = CV = \left(\frac{1 \times 2}{1 + 2} \right) \mu\text{F} \times 100 = \frac{200}{3} \ \mu\text{C} = 66.7 \ \mu\text{C}$$

$$V_1 = \frac{Q}{C_1} = \frac{66.7 \times 10^{-6}}{1 \times 10^{-6}} = 66.7 \text{ V}$$

$$V_2 = \frac{Q}{C_2} = \frac{66.7 \times 10^{-6}}{2 \times 10^{-6}} = 33.3 \text{ V}$$

therefore $V_1 + V_2 = 66.7 + 33.3 = 100$ V.

9. (i) $Q = CV = \dfrac{10 \times 20}{10 + 20} \times 10^{-6} \times 200 = 1.33 \text{ mC}$

(ii) $W = \dfrac{1}{2} CV^2 = \dfrac{1}{2} \times \dfrac{10 \times 20}{10 + 20} \times 10^{-6} \times 200^2 = 0.133 \text{ J}$

(iii) $Q = CV$

$V_2 = 60 \text{ V}$ and $V_1 = 140 \text{ V}$

$Q = C_1 V_1 = 10 \times 10^{-6} \times 140 = 1.4 \text{ mC}$

$Q = C'V_2 = 1.4 \text{ mC}$ where $C' = C + C_2$

$C' = \dfrac{1.4}{60} \text{ mC} = 23.3 \ \mu\text{F}$

$C = C' - C_2 = 23.3 \ \mu\text{F} - 20 \ \mu\text{F} = 3.3 \ \mu\text{F}$

$C = 3.3 \ \mu\text{F}$ is the required value

10. (i) $C_T = 2C \ /\!/ \ C \ /\!/ \ 2C = (2C \ /\!/ \ 2C) \ /\!/ \ C = C \ /\!/ \ C = \dfrac{C}{2}$

(a) $C = 1 \ \mu\text{F}, \quad C_T = 0.5 \ \mu\text{F}$

(b) $C = 10 \ \mu\text{F}, \quad C_T = 5 \ \mu\text{F}$

(ii) $C_T = \dfrac{3C \times \dfrac{C}{3}}{3C + \dfrac{C}{3}} = \dfrac{\dfrac{3C^2}{3}}{\dfrac{10C}{3}} = \dfrac{3}{10} C$

(a) $C = 1 \ \mu\text{F}, \quad C_T = \dfrac{3}{10} \ \mu\text{F}.$

(b) $C = 10 \ \mu\text{F}, \quad C_T = 3 \ \mu\text{F}.$

SOLUTIONS 4

$$\frac{IN}{l} = \frac{\phi}{A\mu_o\mu_r}$$

$$I = \frac{l\phi}{NA\mu_o\mu_r} = \frac{0.5 \times 750 \times 10^{-6}}{300 \times 10 \times 10^{-4} \times 4\pi \times 10^{-7} \times 1000} = 0.9947$$

$I = 1$ A approximately.

(i) $F = IN = 5 \times 500 = 2500$ A

(ii) $H = \dfrac{F}{l} = \dfrac{2500}{0.75} = 3333$ A/m

(iii) $B = \dfrac{\phi}{A} = H\mu_o\mu_r$ but a perspex has $\mu_r = 1$

$B = 3333 \times 4\pi \times 10^{-7} = 0.004189$

$B = 4.19$ mT approximately

(iv) $\phi = BA = 4.19 \times 10^{-3} \times 10 \times 10^{-4} = 4.19\ \mu$Wb.

$$I = \frac{lB}{N\mu_o\mu_r} = \frac{1 \times 1.5}{1000 \times 4\pi \times 10^{-7} \times 1000} = \frac{1.5}{4\pi \times 10^{-1}} = 1.19 \text{ A}$$

$$S = \frac{l_i}{\mu_o\mu_r A} + \frac{l_a}{\mu_o A} = \frac{0.999}{4\pi \times 10^{-7} \times 1000 \times 7.5 \times 10^{-4}} + \frac{0.001}{4\pi \times 10^{-7} \times 7.5 \times 10^{-4}}$$

$$= \frac{(795 + 796)}{7.5 \times 10^{-4}} = 2121000 \text{ A/Wb}$$

$l_i = 1000$ mm $-$ 1 mm $= 999$ mm $= 0.999$ m

$l_a = 0.001$ m

$S = \dfrac{IN}{\phi} = \dfrac{IN}{BA}$ therefore

$$B = \frac{IN}{AS} = \frac{1.19 \times 1000}{7.5 \times 10^{-4} \times 2121000} = 0.74775 \text{ T}$$

4. (a) $S = \dfrac{l}{\mu_o \mu_r A} = 400000 = \dfrac{0.628}{4\pi \times 10^{-7} \mu_r A}$

since $D = 20$ cm, $l = 20\pi = 0.628$ m

$$\mu_r = \dfrac{0.628}{4\pi \times 10^{-7} \times 400000 \times 6 \times 10^{-4}} = 2080 \text{ to 3 S.F.}$$

(b) $N = 250, I = 1$ A,

$$S = \dfrac{l_i}{\mu_o \mu_r A} + \dfrac{l_a}{\mu_o A} = 400000 + 400000 = 800000$$

$S = 800000$ A/Wb

$$S = \dfrac{IN}{\phi}$$

$$\phi = \dfrac{IN}{S} = \dfrac{250 \times 1}{800000} = 313 \ \mu\text{Wb}.$$

5. (a) Fig. 238 shows the B/H curve

Fig. 238 Hysteresis loop

(i)   OC

(ii)   OD

(iii) the initial magnetization curve OA

(iv) B_{sat}

(b) $S_i = \dfrac{l_i}{\mu_o \mu_r A} = \dfrac{0.5}{4\pi \times 10^{-7} \times 10^3 A} = 500000 \text{ A/Wb}$

$S_a = \dfrac{l_a}{\mu_o A} = \dfrac{0.5 \times 10^{-3}}{4\pi \times 10^{-7} A} = 5 \times 10^5$

$A = \dfrac{0.5 \times 10^{-3}}{5 \times 10^5 \, 4\pi 10^{-7}} = \dfrac{0.05}{20\pi} = 7.96 \times 10^{-4} \text{ m}^2$

$S = S_i + S_a = 5 \times 10^5 + 5 \times 10^5 = 1 \times 10^6 \text{ A/Wb}$

$S = \dfrac{IN}{\phi} = 1 \times 10^6 \text{ A/Wb}$

$N = \dfrac{1 \times 10^6 \times 0.5 \times 10^{-3}}{1.5} = \dfrac{500}{1.5} = 333.$

). Data: $l_i = 0.25 \text{ m}, A = 2 \times 10^{-4} \text{ m}^2, N = 1000,$

$l_a = 0.001 \text{ m}, \phi = 100 \times 10^{-6} \text{ Wb}$

$S = \dfrac{l_i}{\mu_o \mu_r A} + \dfrac{l}{\mu_o A} = \dfrac{0.25}{4\pi \times 10^{-7} \mu_r 2 \times 10^{-4}} + \dfrac{0.001}{4\pi \times 10^{-7} \times 2 \times 10^{-4}}$

where μ_r is found below

$B = 0.5$ T from the magnetisation curve graph corresponding to a value
$H = 2250$ A/m

$B = \dfrac{\phi}{A} = \dfrac{100 \times 10^{-6}}{2 \times 10^{-4}} = 0.5 \text{ T}$

$\mu_o \mu_r = \dfrac{B}{H} = \dfrac{0.5}{2250}$

$\mu_r = \dfrac{0.5}{2250 \times 4\pi \times 10^{-7}} = 177$

$S = \dfrac{0.25}{4\pi \times 10^{-7} \times 177 \times 2 \times 10^{-4}} + \dfrac{0.001}{4\pi \times 10^{-7} \times 2 \times 10^{-4}}$

$= 5620000 + 3980000 = 9600000$

$S = 9600000 \text{ A/Wb}$

$$S = \frac{IN}{100 \times 10^{-6}} = \frac{I \times 1000}{100 \times 10^{-6}} = 9600000 \text{ A/Wb}$$

$I = 0.96$ A

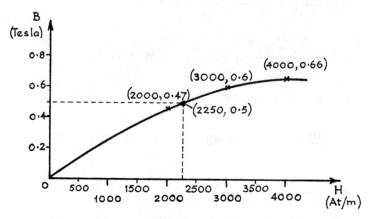

Magnetisation Curve for the specimen.

Fig. 239

7. (i) $R = \varrho \dfrac{l}{A}$

where R is the resistance of copper wire in ohms, ϱ is the resistivity of the copper wire Ωm, l is the length of the copper wire in m, and A is the cross sectional area in m^2.

The analogous expression in a magnetic circuit is that of reluctance S.

$$S = \left[\frac{1}{\mu}\right] \frac{1}{A}$$

where l is the length of the magnetic path and A is the cross sectional area of the ferromagnetic material where $\dfrac{1}{\mu}$ is analogous to ϱ or μ is analogous to σ, the conductivity.

The units of S are A/Wb, of l is m, of A is m^2, and μ is henrys per metre.

(ii) $F = NI = \phi S = \phi \dfrac{l}{\mu_o \mu_r A}$ where S is the reluctance and

$\mu = \mu_o \mu_r$ = absolute permeability.

198

The analogous expression in an electric circuit is that of ohm's law: e.m.f. = current × resistance $E = IR$ current is therefore analogous to flux and e.m.f. is analogous to magnetomotive force or ampere-turns in a magnetic circuit and reluctance is analogous to resistance.

8. (i) $S = \dfrac{l}{\mu_o \mu_r A} = \dfrac{400 \times 10^{-3}}{4\pi \times 10^{-4} \times 400 \times 10^{-6}} = \dfrac{10^7}{4\pi} = 795775 \text{ A/Wb}$

    (ii) $\phi = \dfrac{NI}{S} = \dfrac{100 \times 5}{795775} = 628 \ \mu\text{Wb}.$

9. The reluctance of the air gap is given by

$$S = \frac{l}{\mu_o A}$$

where $\mu_r = 1$

$S = \dfrac{4 \times 10^{-3}}{4\pi \times 10^{-7} \times 400 \times 10^{-6}} = 7957747$

$= 7957747 \text{ A/Wb}.$

The total reluctance

$S = S_i + S_a$

S_i = the reluctance of the iron

S_a = the reluctance of the air gap.

$l_i = 400 - 4 = 396 \text{ mm}$

$S_i = 795775 \times \dfrac{400}{396} = 803813$

$S = 803813 + 7957747 = 8761560 \text{ A/Wb}.$

The total reluctance.

The effect of introducing the air gap is to increase the reluctance of the circuit. The reluctance of the magnetic circuit with the gap is nearly 11 times larger than the reluctance without the gap and hence the magnetic flux is 11 times smaller

$$\phi = \frac{IN}{S} = \frac{5 \times 100}{8761560} = 57 \ \mu\text{Wb}.$$

SOLUTIONS 5

1. $F = BIl$

    where F = force in newtons, B is the flux density in teslas, I is the current in amperes and l = the length of a conductor in metres.

    $$B = \frac{F}{Il} = \frac{0.25}{0.05 \times 0.5} = \frac{0.250}{0.025} = 10 \text{ T}.$$

2. $F = BIl \sin \theta$

    $$= 0.9 \times 0.15 \times 0.4 \times \sin 80° = 0.053 \text{ N}.$$

    The force is now $\left(\dfrac{1}{5}\right) \times 0.053 = 0.0106 \text{ N}$

    $$0.0106 = 0.9 \times 0.15 \times 0.4 \sin \theta$$

    $$\sin \theta = \frac{0.0106}{0.9 \times 0.15 \times 0.4} = 0.1962963$$

    $$\theta = 11.32°.$$

3. $F = BIl$

    $$I = \frac{F}{Bl} = \frac{100 \times 10^{-3}}{50 \times 10^{-3} \times 10 \times 10^{-2}} = 20 \text{ A}.$$

4. $v = \omega r$ is the relationship connecting angular and linear velocities. The radius of rotation, $r = \dfrac{15}{2} = 7.5 \text{ cm}$, $\omega = \dfrac{2\pi}{60} \times 500 = 53.36 \text{ radians per second}$.

    $v = 52.36 \times 0.075 = 3.93 \text{ m/s}.$

    Fig. 240 shows the rectangular coil

Fig. 240 Rectangular coil.

$E = Blv = 1 \times 0.2 \times 3.93 = 0.79$ volts

For 100 turns, $E = 0.79 \times 100 = 79$ volts.

5. $E = N\dfrac{d\phi}{dt}$

$N = \dfrac{E}{d\phi/dt} = \dfrac{5}{0.5} = 10.$

6. $E = L\dfrac{di}{dt} = 200 \times 10^{-3} \times 500 = 100$ V.

7. Two coils of self inductance L_A and L_B are placed close to each other as shown in Fig. 241. When the current in L_A changes then the e.m.f. induced across the load resistor, R is $E_2 = M\dfrac{di}{dt}$, that is, an induced e.m.f. across R is established where there is a change of current in L_A.

Fig. 241 Mutual inductance for coils close to each other.

When the coils are not placed close to each other, M may be negligible and the e.m.f. induced across R is also negligible. The concept of mutual inductance is well illustrated.

$$E_2 = M\frac{di}{dt} = 0.2 \times 100 = 20 \text{ V.}$$

8. $\dfrac{E_1}{E_2} = \dfrac{N_1}{N_2} = \dfrac{I_2}{I_1}$ and $R_{in} = \left(\dfrac{N_1}{N_2}\right)^2 R$

$\dfrac{N_2}{N_1} = 10,\ E_1 = 240$ V and $R = 12$ KΩ.

(i) $E_2 = \dfrac{E_1}{N_1/N_2} = \dfrac{240}{1/10} = 2400$ V note that the transformer is a step up.

$I_2 = \dfrac{E_2}{R} = \dfrac{2400}{12000} = 0.2$ A.

(ii) $I_1 = \dfrac{I_2}{N_1/N_2} = \dfrac{0.2}{1/10} = 2$ A.

(iii) $R_{in} = \left(\dfrac{N_1}{N_2}\right)^2 R = \left(\dfrac{1}{10}\right)^2 1200 = 120\ \Omega.$

9.   (i) $LI = N\phi \qquad\qquad H = \dfrac{IN}{l}$

$L = \dfrac{N\phi}{I} = N\dfrac{BA}{Hl/N} = N^2\dfrac{BA}{Hl} = \dfrac{N^2\mu_o\mu_r A}{l}$

$L = \dfrac{N^2\mu_o\mu_r A}{l}$

L depends greatly on the number of turns and also depends on μ_r, the relative permeability of the magnetic material.

(ii) $S = \dfrac{l}{\mu_o A}$ reluctance of a non-magnetic material the reluctance is

independent of the I and since S depends on l, μ_o and A which are

constants, hence it is constant.

(iii) $S = \dfrac{1}{\mu_o\mu_r A}$ reluctance of magnetic material

$\dfrac{B}{H} = \mu_o\mu_r = \dfrac{\phi/A}{IN/l} = \dfrac{\phi l}{AIN}$ therefore

$\mu_r = \dfrac{\phi l}{AIN\mu}$ and hence it is inversely proportional to I.

10.   See text and Fig. 131.

SOLUTIONS 6

1.   (i)   A non-sinusoidal wave is shown in Fig. 242

Fig. 242 A non-sinusoidal waveform.

(ii)   A sinusoidal current wave is shown in Fig. 243

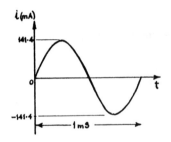

Fig. 243 A sinusoidal waveform.

$$I_{max} = 100 \times \sqrt{2} \text{ mA} = 141.4 \text{ mA}$$

$$T = \frac{1}{1000} = 1 \text{ ms.}$$

(iii) A sinusoidal voltage wave is shown in Fig. 244

Fig. 244 A lagging sinusoidal voltage wave

203

$$T = \frac{1}{f} = \frac{1}{50} = \frac{1000}{50} \text{ ms} = 20 \text{ ms}$$

$30° = \dfrac{\pi}{6}$ phase angle lagging

2. $v_1 = 15 \sin \omega t$ is a reference and it is drawn horizontally $AB = 15$ V, $BC = 25$ V. Using the cosine rule to the triangle ABC

$$AC^2 = AB^2 + BC^2 - 2 \times AB \times BC \times \cos 135°$$
$$= 15^2 + 25^2 - 2 \times 15 \times 25(- 0.707)$$
$$= 225 + 625 + 530.25 = 1380.25$$

$$AC = \sqrt{1380.25} = 37.15 \text{ V} = \text{is the resultant.}$$

The phase angle of the resultant, ϕ is found by the sine rule

$$\frac{25}{\sin \phi} = \frac{37.15}{\sin 135°}$$

$$\sin \phi = \frac{25 \times \sin 135°}{37.15} = 0.4758457$$

$$\phi = 28.4°$$

$$v_R = v_1 + v_2 = 37.15 \sin \left(\omega t - 28.4°\right)$$

$$v_R = 37.15 \sin \left(\omega t - 28.4°\right)$$

$$28.4° = 28.4 \times \frac{\pi}{180} = 0.496^c$$

therefore $v_R = 37.2 \sin \left(\omega t - 0.496^c\right)$.

To phasor diagram of Fig. 245 is drawn to scale by using a ruler, a set square and a protractor. Use 10 V = 1 cm scale. Measure AC and the angle ϕ.

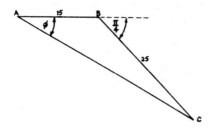

Fig. 145 Phasor

3. (i) $V_{max} = 100\sqrt{2} = 141.4$ V

(ii) $V_{av} = \dfrac{2}{\pi} \times 100\sqrt{2} = 90$ V

(iii) $V_{r.m.s.} = \dfrac{100\sqrt{2}}{\sqrt{2}} = 100$ V

(iv) $\omega = 377$ radians per second

(v) $f = \dfrac{377}{2\pi} = 60$ Hz

(vi) $T = \dfrac{1}{60} = 16.7$ ms

(vii) $v = 141.4\sin\left(377 \times 5 \times 10^{-3} + \dfrac{\pi}{4}\right)$

$= 141.4\sin\left(1.885^c + 0.7854^c\right) = 141.4\sin 2.67^c$

$= 64.2$ V

(viii) $\phi = \dfrac{\pi}{4}$ leading.

4. $i = \dfrac{v}{100} = \dfrac{10}{100}\sin\left(500\pi t - \dfrac{\pi}{6}\right)$

$i = 100\sin\left(500\pi t - \dfrac{\pi}{6}\right)$ mA

205

Fig. 246 shows the waveforms of i and v which are in phase.

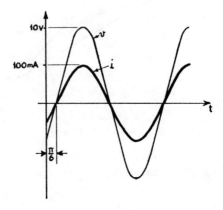

Fig. 246 Waveforms in phase.

5. (a) (i) $f = \dfrac{314200}{2\pi} = 50$ KHz

   (ii) $I = \dfrac{100 \times 10^{-3}}{\sqrt{2}} = 70.7$ mA

   (iii) $i = 100 \times 10^{-3} \sin\left(314200 \times 2 \times 10^{-6} - \dfrac{\pi}{12}\right)$

   $= 100 \times 10^{-3} \sin\left(0.6284^c - 0.2617993^c\right)$

   $= 100 \times 10^{-3} \sin 0.3666^c = 100 \times 10^{-3} \times 0.358$

   $= 35.8$ mA.

(b) Power developed in the 2.2 KΩ $= I^2 R$ where I is the r.m.s. value of current

$$P = \left(\dfrac{100 \times 10^{-3}}{\sqrt{2}}\right)^2 \times 2200 = 11 \text{ W.}$$

(c) Fig. 247 shows that the current lags the voltage by $\dfrac{\pi}{12}$ or 15°.

206

Fig. 247 Phasor diagram.

6. Referring to Fig. 248

E_1 is drawn horizontally representing 10 V = 1 cm, $OA = E_1$, $AB = E_2$, drawn at an angle of 60° leading to the horizontal, $BC = E_3$, drawn at an angle of 90° lagging to the horizontal. The resultant voltage is given by E = 135 V since OC = 13.5 cm and its angle to the horizontal is about 3°.

$$E_R = E_M \sin\left(\omega t + 3°\right) \text{ or}$$

$$E_R = 135 \sin\left(\omega t + 0.052^c\right).$$

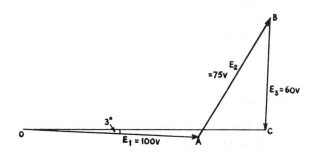

Fig. 248

7. $V_{\text{r.m.s.}} = 100\sqrt{2}$ volts

$f = 50$ Hz

$v = V_{\text{max}} \sin \omega t$

$v = 100\sqrt{2}\sqrt{2} \sin 2\pi ft = 200 \sin\left(2\pi \times 50t\right) = 100$

$$\sin(2\pi \times 50t) = \frac{100}{200} = \frac{1}{2} = \sin\frac{\pi}{6}.$$

$$2\pi \times 50t = \frac{\pi}{6}$$

$$t = \frac{1}{600} \times 1000 \text{ ms}$$

$$t = 1.67 \text{ ms.}$$

8. $i = 200\sqrt{2} \sin 877t$ mA

(i) $I = \dfrac{200\sqrt{2} \times 10^{-3}}{\sqrt{2}} = 200$ mA $= 0.2$ A

(ii) $2\pi ft = 877t$

$$f = \frac{877}{2\pi} = 139.6 \text{ Hz}$$

(iii) $i = 200\sqrt{2} \sin 877 \times 10^{-3}$

$$= 200\sqrt{2} \sin 0.877^{c}$$

$$= 217.43 \text{ mA} = 0.22 \text{ A.}$$

9. Average power dissipated

(i) by calculation

$$P = I^2 R = IV = \frac{V^2}{R}$$

$$= 10\sqrt{2} \times 220 = 3110.8 \text{ W.}$$

(ii) from the waveform of power see graph in text of Fig. 147.

10. $v = 500 \sin \left(2\pi \times \dfrac{1}{10 \times 10^{-3}} t + \dfrac{\pi}{10} \right)$

$v = 500 \sin \left(200\pi t + \dfrac{\pi}{10} \right)$ V.

SOLUTIONS 7

1. $X_C = \dfrac{1}{2\pi f C} = \dfrac{1}{2\pi \times 100 \times 100 \times 10^{-6}} = 15.9\ \Omega.$

2. $X_L = 2\pi f L = 2\pi \times 60 \times 50 \times 10^{-3} = 18.9\ \Omega.$

3. $X_C = \dfrac{V}{I}$ or $I = \dfrac{V}{X_C} = \dfrac{50}{100} = 0.5$ A.

4. $X_L = \dfrac{V}{I}$ or $V = I X_L = 5 \times 5 = 25$ V.

5. Fig. 249 shows the phase relationships for i, v_L and v_C.

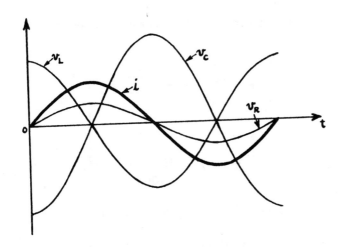

Fig. 249 Waveforms out of phase.

6. Fig. 250 shows the circuit.

Fig. 250 *L-R* series circuit.

209

(i) $Z = \sqrt{R^2 + X_L^2} = \sqrt{50^2 + 157.1^2} = 164.9 \ \Omega.$

where $X_L = 2\pi \times 50 \times 0.5 = 157.1 \ \Omega$

(ii) $I = \dfrac{240}{164.9} = 1.46 \ A$

(iii) $V_L = IX_L = 1.46 \times 157.1 = 229.3 \ V$

(iv) $V_R = IR = 1.46 \times 50 = 73 \ V$

$V^2 = V_R^2 + V_L^2 = 73^2 + 229.3^2$

$V = \sqrt{73^2 + 229.3^2} = 240 \ V.$

7.   Fig. 251 and Fig. 252 show the d.c. and a.c. circuit of LR in series.

Fig. 251 *LR* d.c. circuit.

The 100 V is now across R since L is fully magnetised, $R = \dfrac{100}{5} = 20 \ \Omega.$
The resistance of the coil is 20 Ω.

Fig. 252 *LR* a.c. circuit.

$$I = \frac{V}{Z} = \frac{100}{Z} = 1 \text{ A}$$

$$Z = 100 \ \Omega.$$

the impedance of the coil at 50 Hz is 100 Ω.

8.   (i) $Z = \sqrt{R^2 + X_L^2} = \frac{V}{I} = \frac{200}{10} = 20 \ \Omega$

  (ii) $P = IV \cos \phi$

$$\cos \phi = \frac{P}{IV} = \frac{500}{10 \times 200} = 0.25$$

  (iii)   From the impedance triangle

$$\cos \phi = \frac{R}{Z}$$

$$R = Z \cos \phi = 20 \times 0.25 = 5 \ \Omega.$$

  (iv) $Z^2 = R^2 + X_L^2$

$$X_L^2 = Z^2 - R^2$$

$$X_L = \sqrt{Z^2 - R^2} = \sqrt{20^2 - 5^2} = \sqrt{400 - 25} = \sqrt{375} = 19.36 \ \Omega$$

$$2\pi f L = 19.36$$

$$L = \frac{19.36}{2\pi 50} = 61.6 \text{ mH.}$$

9.   (i) $P = IV \cos \phi$

$$\cos \phi = \frac{P}{IV} = \frac{1000}{16 \times 240} = 0.26$$

  (ii) $Z = \frac{V}{I} = \frac{240}{16} = 15 \ \Omega$

  (iii)   From the equation $\cos \phi = \frac{R}{Z}$

$$R = Z \cos \phi = 15 \times 0.26 = 3.9 \ \Omega$$

(iv) $X_L = \sqrt{Z^2 - R^2} = \sqrt{15^2 - 3.9^2} = 14.48 \ \Omega$

(v) $X_L = 2\pi fL$

$$L = \frac{X_L}{2\pi f} = \frac{14.48}{2\pi \times 50} = 46.1 \text{ mH}.$$

10. Fig. 253 shows the *LRC* series circuit.

Fig. 253 Series *LRC* circuit at resonance.

(i) $f_o = \dfrac{1}{2\pi \sqrt{LC}}$

$= \dfrac{1}{2\pi \sqrt{300 \times 10^{-3} \times 100 \times 10^{-6}}}$

$= \dfrac{1000}{2\pi \sqrt{30}} = 29.06 \text{ Hz}$

(ii) $I_o = \dfrac{240}{15} = 16 \text{ A}$

(iii) $V_L = I_o \times X_L = 16 \times 2\pi \times 29.06 \times 300 \times 10^{-3} = 876.4 \text{ V}$

(iv) $V_C = I_o X_C = 16 \times \dfrac{1}{2\pi \times 29.06 \times 100 \times 10^{-6}} = 876.4 \text{ V}$

(v) $P = I_o^2 R = 16^2 \times 15 = 3840 \text{ W}.$

11. The maximum current is the condition of resonance.
Fig. 254 shows the circuit.

212

Fig. 254 Resonant series circuit.

$$R = \frac{V}{I_o} = \frac{240}{1} = 240 \ \Omega$$

$X_L = X_C$ at resonance

$$f_o = \frac{1}{2\pi\sqrt{LC}} \text{ squaring up both sides}$$

$$LC = \frac{1}{4\pi^2 f_o^2} = \frac{1}{98696.044} = 1.0132118 \times 10^{-5}$$

$$V_C = I_o X_C = 1 \times \frac{1}{2\pi f_o C} = 500 \text{ hence } C = \frac{1}{2\pi \times 50 \times 500} = 6.37 \ \mu F$$

$$\text{and } L = \frac{1.0132118 \times 10^{-5}}{6.37 \times 10^{-6}} = 1.59 \text{ H.}$$

12.   Fig. 255 shows the LRC series circuit.

Fig. 255 LRC series resonant circuit.

$$\text{(i)} \quad f_o = \frac{1}{2\pi\sqrt{LC}} = \frac{1}{2\pi\sqrt{0.05 \times 0.2 \times 10^{-6}}} = 1591.6 \text{ Hz}$$

(ii) $I_o = \dfrac{300}{50} = 6\text{ A}$

(iii) $V_C = I_o X_C = 6 \times \dfrac{1}{2\pi \times 1591.6 \times 0.2 \times 10^{-6}} = 2999.91$

   $V_C = 3000$ volts approximately

(iv) $V_C = QV$

   $Q = \dfrac{V_C}{V} = $ voltage amplification factor or quality factor

   $= \dfrac{3000}{300} = 10$

   $Q = 10.$

13. Fig. 256 and Fig. 257 show the circuit and the impedance triangle respectively.

Fig. 256 *LRC* series circuit.

Fig. 257 Impedance triangle.

214

$$Z = \sqrt{R^2 + (X_C - X_L)^2} = \frac{240}{5} = 48 \ \Omega$$

$$X_L = 2\pi f L = 2\pi \times 50 \times 50 \times 10^{-3} = 15.7 \ \Omega$$

$$(X_C - X_L)^2 = 48^2 - R^2$$

$$X_C - X_L = \sqrt{48^2 - 10^2} = 47 \ \Omega$$

$$X_C = 47 + 15.7 = 62.7 \ \Omega$$

$$\frac{1}{2\pi 50 C} = 62.7$$

$$C = \frac{1}{2\pi \times 50 \times 62.7} = 51 \ \mu F$$

$$Z = \sqrt{R^2 + (X_L - X_C)^2} = \frac{240}{2} = 120 \ \Omega$$

$$X_L = X_C = \sqrt{120^2 - 10^2} = 119.6 \ \Omega$$

$$X_L - 119.6 = X_C$$

$$X_C = 15.7 - 119.6 = 103.9 \ \Omega$$

$$X_C = \frac{1}{2\pi f C} = 103.9$$

$$C = \frac{1}{2\pi \times 50 \times 103.9} = 30.6 \ \mu F.$$

SOLUTIONS 8

1. See in the text constructional details of a moving coil instrument.

2. Fig. 258 shows the shunt circuit.

Fig. 258 Shunt circuit Ammeter.

(i) Let R_{sh} be the resistor required across the 15 mA moving coil milliammeter.

If the total current is 5 A then the current through R_{sh} is 5 – 0.015 = 4.985 A.

The p.d. across the 10 Ω moving coil = 0.015 × 10 = 0.15 V.

The p.d. across R_{sh} = 4.985 × R_{sh} = 0.15

$$R_{sh} = \frac{0.15}{4.985} = 0.03 \ \Omega.$$

(ii) Fig. 259 shows the multiplier circuit.

Fig. 259 Multiplier circuit Voltmeter.

$500 = 0.15 + 0.015 R_M$

$0.015 R_M = 499.85$

$$R_M = \frac{499.85}{0.015} = 33323.33 \ \Omega.$$

216

3. Fig. 260 shows the circuit

Fig. 260 Shunt circuit.

The p.d. across AB = $2 \times 10^{-3} \times (400 + 600)$

= 2 V

$$R_{sh} = \frac{2}{0.998} = 2.004 \ \Omega.$$

4. See text

$$R_x = \left[\frac{R_1}{R_2} \right] R_3 = \text{arm's ratio} \times R_3 .$$

5. Fig. 261 shows the multiplier circuit.

Fig. 261 Multiplier circuit.

$$I = \frac{50}{500000} = 10^{-4} \ A$$

$$10^{-4} R_M = 450$$

$$R_M = \frac{450}{10^{-4}} = 450 \times 10^4 = 4.5 \ M\Omega.$$

6. (i) At balance

$$500 \times 25 = 100 \times R_x$$

$$R_x = \frac{500 \times 25}{100} = 125 \; \Omega.$$

(ii) Current through the 100 Ω resistor $= \dfrac{6}{125} = 48$ mA

p.d. across the 100 $\Omega = 48 \times 10^{-3} \times 100 = 4.8$ V.

Alternatively

Using the potential divider method p.d. across the 100 $\Omega = \dfrac{100}{125} \times 6$

$$= 4.8 \; \text{V}.$$

(iii) the current through the 500 $\Omega = \dfrac{6}{500 + R_x} = \dfrac{6}{500 + 125}$

$$= \dfrac{6}{625} = 9.6 \; \text{mA}.$$

7. See text for potential divider at balance, no current flows through G.

$$R_3 R_1 = R_2 R_x$$

$$R_x = \left[\frac{R_1}{R_2} \right] R_3$$

Fig. 262 shows two potential dividers.

Fig. 262 Wheatstone bridge.

The wheatstone bridge method for measuring a resistance is fairly accurate. The ammeter and voltmeter method for measuring a resistance is rather inaccurate since the p.d. across an ammeter has to be known and taken into account, and similarly the current through the voltmeter should be known and taken into account in calculating accurately the value of resistance.

Applying the current divider method of Fig. 263, we have

Fig. 263 Shunt circuit.

$$I_1 = \frac{0.015}{0.05 + 0.015} I$$

$$5 = \frac{0.015}{0.065} I$$

$$I = \frac{5 \times 0.065}{0.015} = 21.7 \text{ A.}$$

0. Fig. 264 shows the potential divider

Fig. 264 Potential divider.

219

If the voltmeter had infinite resistance, the p.d. indicated would have been 100 volts. The p.d. across the voltmeter is found by using the potential divider

$$V = \frac{500}{1500} \times 200 = \frac{200}{3} = 66.7 \text{ V}$$

$$V = 66.7 \text{ V}.$$

11. Again if the voltmeter were ideal, the p.d. across R_1 is found by the potential divider

$$V = \frac{1000}{1470} \times 100 = 68.03 \text{ volts}.$$

The actual p.d. across R_1 is found by lumping the two parallel resistors of 1000 Ω and applying the potential divider

$$V = \frac{500}{970} \times 100 = 51.55 \text{ volts}.$$

The error in the calculated value is $68.03 - 51.55 = 6.48$ volts.

12. (i) From Fig. 265, $R = \dfrac{5}{10 \text{ mA}} = \dfrac{5000}{10} = 500 \text{ Ω}$

Fig. 265 Moving coil instrument.

(ii) From Fig. 266, $38 = 5 + 10 \times 10^{-3} R_M$

$$R_M \, 10 \times 10^{-3} = 38 - 5$$

$$R_M = \frac{33}{10 \times 10^{-3}} = 3.3 \times 10^3 \text{ Ω}$$

$$R_M = 3.3 \text{ KΩ}.$$

Fig. 266 Multiplier.

(iii) From Fig. 267, $R_{sh}\, 9.99 = 5$

$$R_{sh} = \frac{5}{9.99} = 0.5005\ \Omega.$$

Fig. 267 Shunt.

3. Fig. 268 shows the basic circuit of an ohmmeter.

Fig. 268 Ohmmeter.

Short circuit the terminals X and Y, the total resistance of the circuit will be:

$$R = \frac{15}{15 \times 10^{-3}} = 1\ K\Omega$$

221

(i)   If I = 1 mA, $R = \dfrac{15}{1 \times 10^{-3}}$ = 15 KΩ

(ii) If I = 5 mA, $R = \dfrac{15}{5 \times 10^{-3}}$ = 3 KΩ

(iii) If I = 10 mA, $R = \dfrac{15}{10 \times 10^{-3}}$ = 1.5 KΩ

In (i)   (15 – 1) KΩ = 14 KΩ

In (ii)   (3 – 1) KΩ = 2 KΩ

In (iii) (1.5 – 1) KΩ = 0.5 KΩ.

14. (i)   Set the amplifier setting to 5 V/cm and the time base setting to 10 ms/cm

     (ii) Set the amplifier setting to 10 mV/cm and the time base setting 10 μs/cm

     (iii) Set the time base setting to 10 μs/cm and a suitable amplifier setting o
           5 V/cm or 1 V/cm.

MISCELLANEOUS

1. Using Kirchhoff's laws, calculate

   (a) the three branch currents in the circuit of Fig. 1,

   (b) the p.d. between the points X and Y stating which point is at the higher potential.

Fig. 1

(Ans. 1.25 A, 1.00 A, 0.25, V_{XY} = 110 V).

2. (a) State Kirchhoff's current and voltage laws.

   (b)

Fig. 2

Determine the value of each branch current in the above circuit.

(Ans. 0.625 A, 0.438 A, 1.06 A).

3. Use Kirchhoff's laws to obtain the currents I_1, I_2 and I_3 in Fig. 3.

Fig. 3

(Ans. $I_1 = 7.5$ A, $I_2 = 5.5$ A, $I_3 = 2$ A)

4.   The circuit shown in Fig. 4 which, includes the internal resistances of the batteries, has a load resistance connected across terminals X and Y.

   The current that flows in this load is 2.25 amperes and the power developed is 13.5 watts.

   Calculate by any method (i)   the current in each branch;
   (ii) the internal resistance (R) of battery A.

Fig. 4

(Ans. 0.75 A, 1.5 A, 8 Ω)

5.   Six resistors are connected as shown in Fig. 5.

   If a battery having an internal resistance of 1 Ω and an e.m.f. of 24 V is connected to terminals A and B, calculate:

   (i)   the current from the battery;
   (ii)   the p.d. across the 8 Ω and 4 Ω resistors;
   (iii) the current taken from the battery if a conductor of negligible resistance is connected in parallel with the 8 Ω resistor.

224

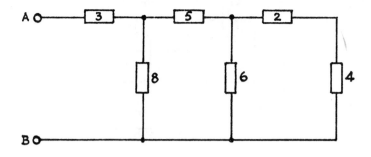

Fig. 5

(Ans. (i) 3 A (ii) 12 V, 3 V (iii) 6 A)

6. With reference to Fig. 6, determine the magnitude and direction of current in each resistor. Use Kirchhoff law.

Fig. 6

(Ans. 0.2 A, 4.18 A, 4.38 A)

7. Two capacitors 9 μF and 18 μF are connected in series. Across both of these is connected a capacitor of 4 μF. The whole combination is connected across a 12 V d.c. supply.

Calculate: (a)   the total circuit capacitance;
         (b)   the p.d. across the 9 μF capacitor;
         (c)   the charge on the 18 μF capacitor;
         (d)   the energy stored by the 4 μF capacitor.

(Ans. (a) 10 μF (b) 8 V (c) 72 μC (d) 288 μJ).

8. In the circuit of Fig. 7, $C_1 = 30$ μF, the energy stored in C_3 is 0.081 J and the p.d. across C_3 is 45 V.

Calculate: (a)   the capacitance of C_3
         (b)   the capacitance of C_2

225

Fig. 7

(Ans. (a) 80 μF (b) 18 μF)

9.   A 0.1 μF capacitor charged to a p.d. of 1000 V is connected across a second uncharged capacitor of 0.4 μF, the source of voltage being removed before the second capacitor is added to the first.

Find (a)   the resulting p.d. across the combined capacitors

   (b)   the energy stored in each before and after paralleling. Comment on the results obtained.

   (c)   If the two capacitors were discharged and then connected in series across the 1000 V supply, what would be the voltage across each capacitor?

(Ans. (a) 200 V (b) 0.05 J, 0 J, 0.002 J, 0.008 J (c) 200 V, 800 V).

10.   A capacitor is formed by two flat metal plates, each of area 5000 mm^2 and separated by a dielectric 1.0 mm thick. The capacitance of this arrangement is 200 pF and a p.d. of 10 KV is applied to the terminals.

$$\in_o \; = \; 8.85 \times 10^{-12} \; F/m$$

Calculate: (i)   the charge on the plates;

   (ii)   the relative permittivity of the dielectric;

   (iii) the electric flux density;

   (iv)   the energy stored.

(Ans. (i) 2 μC (ii) 4.52 (iii) 4 × 10^{-3} C/m^2 (iv) 0.01 J)

11.   Two capacitors A and B have capacitances of 8 μF and 16 μF respectively.

Calculate the voltage of A and the energy stored in A when A and B are connected (i)   in series, and

   (ii)   in parallel

with a d.c. supply of 12 V in both cases.

(Ans. (i) 8 V, 256 μJ (ii) 12 V, 576 μJ.)

12.   In the circuit of Fig. 8, $C_1 = 60 \; \mu$F

   $C_2 = 90 \; \mu$F

and the total capacitance is 100 μF.

226

Fig. 8

Calculate   (i)   the capacitance of C_3
            (ii)   the voltage across of C_3
            (iii) the charge on C_2
            (iv) the total energy supplied.

(Ans. (i) 300 μF (ii) $\dfrac{100}{3}$ V (iii) 6 mC (iv) $\dfrac{1}{2}$ J)

3.   (a)   On what factors does the capacitance of a capacitor depend?

    (b)   For the circuit shown in Fig. 9 calculate
         (i)   the equivalent capacitance
         (ii)   the total energy stored
         (iii) the voltage across the 10 μF.

Fig. 9

(Ans. (i) 10 μF (ii) 0.2 J (iii) 100 V).

227

14. (a) A parallel plate capacitor has a capacitance of 200 pF. What will be the new value of capacitance if the effective plate area is reduced by 25% and the dielectric thickness is doubled?

(b) Calculate the total capacitance if a 40 μF capacitor and a 60 μF capacitor are connected:

(i) in series

(ii) in parallel.

(c) Calculate the p.d. across an 80 μF capacitor when the energy stored by the capacitor is 144 mJ.

(Ans. (a) 75 pF (b) (i) 24 μF, 100 μF (c) 60 volts.)

15. (a) What factors determine the capacitance existing between 2 parallel metal plates?

Give a formula from which this capacitance can be calculated.

(b) Three capacitors 4 μF, 6 μF and 12 μF capacitance are connected in series to a 200 V d.c. supply.

Calculate: (i) the charge on each capacitor

(ii) the p.d. across each capacitor

(iii) the energy stored in each capacitor.

(Ans. (b) (i) 400×10^{-6} C (ii) 100 V, 60.7 V, 33.3 V

(iii) 0.02 J, 0.0133 J, 0.0067 J.)

16. (a) For a given capacitor write down the expressions stating the units employed relating:

(i) the potential difference and charge;

(ii) the potential difference and stored energy.

$(Q = CV, W = \frac{1}{2}CV^2)$

(b) Two air-spaced metal plates insulated from each other, are mounted so as to form a parallel plate capacitor. When the distance between the plates is 2 mm the capacitance is 0.5 pF. Determine the capacitance when the spacing is increase to 5 mm.

How will the value of the capacitor be affected if the whole is immersed in oil having a relative permittivity of 3?

(Ans. 1.5 pF, 0.6 pF)

17. Refer to Fig. 10,

    Calculate (a)   the charge on each capacitor before A is connected,

              (b)   the total energy stored and

              (c)   the value of A which when connected in parallel with the 10 μF capacitor would change V_2 to 80 V with the same supply.

Fig. 10

(Ans. (a) 1.44 mC (b) 0.1728 J (c) 20 μF)

18. A coil having 1500 turns is wound on an iron core having a mean length of 60 cm and a cross-sectional area of 5 cm². Calculate the current required to produce a flux-density of 1.4 T in the core.

What would be the value of the flux-density if there was an equivalent of a 1 mm air gap in the ring and the current remained the same?
(Take μ_r = 1000).

(Ans. 0.45 A, 0.53 T)

19. A ferromagnetic toroid has a mean circumference of 20 cm a cross-sectional area of 1.5 cm² and is wound with a coil of 800 turns.

A 1 mm radial air gap is cut in the toroid.

Calculate the coil current required to establish a flux of 90 μWb in the toroid.

An extract from the B/H characteristic is shown below:

B	tesla	0.47	0.60	0.66
H	amperes/metre	2000	3000	4000

229

20. (a) Define the term 'relative permeability'.

Calculate the relative permeability of a magnetic material having a flux density of 0.6 T when the magnetizing force is 318 A/m.

(b) A ring of magnetic material is uniformly wound with 200 turns. The mean diameter of the ring is 10 cm and its cross-sectional area is 1 cm².

Calculate the flux density in the ring when the relative permeability of the material of the ring is 2500 and the current in the winding is 0.8 A.

(Ans. μ_r = 1502, 1.6 T)

21. A coil having 1000 turns is wound uniformly on an iron ring having a mean circumference of 50 cm and cross-sectional area of 4 cm². An air gap is cut in the ring of such a length that the reluctance of the air gap is equal to the reluctance of the iron ring.

Calculate the current required in the coil to set up a flux density of 1.2 T in the ring and air gap.

Take μ_r = 1500 and μ_o = $4\pi\, 10^{-7}$ H/m

22. A 2000 turn coil is wound on an iron ring having a cross-sectional area of 10 cm², and mean diameter of 0.2 m. There is a radial saw cut 2 mm wide in the ring.

Calculate the current in the coil for a flux of 0.8 mWb in the ring. The B/H curve for the iron is as follows:

H (A/m)	100	200	300	400	500	600
B (T)	0.46	0.78	0.97	1.09	1.16	1.20

State any assumption made.

(Ans. 0.69 A)

23. The magnetic circuit shown is built up of iron of square cross-section 2 cm side. Each of the coils is wound with 1200 turns and the exciting current is 0.8 A. At that current the relative permeability of part A may be taken to be 900 and of part B 1250.

Fig. 11

Find (a)   the reluctance of *A*
     (b)   the reluctance of *B*
     (c)   the reluctance of the two air gaps
     (d)   the total reluctance
     (e)   the m.m.f.
     (f)   the total flux
     (g)   the flux density.

(Ans. 44.22×10^4 A/Wb, 63.68×10^4 A/Wb, 796×10^4 A/Wb,
         903.9×10^4 A/Wb, 1920 A, 212.4 μWb, 0.531 T)

24. (a)   State Lenz's law.

     (b)   A conductor 30 cm in length and carrying a current of 100 mA is placed
            at right angles to a magnetic field having a flux density of 0.8 T.
            Calculate the force on the conductor.

     (c)   An ideal transformer having a 4:1 step-down ratio, has a load current
            and voltage of 100 mA and 60 V respectively. Calculate the primary
            current and voltage.

     (d)   Calculate the energy contained in an inductance of 2 H carrying a
            current of 500 mA.

     (Ans. (b) 0.024 N (c) I_1 = 0.025 A, E_1 = 240 V (d) W = 0.25 J)

25. (a)   Explain what is meant by the terms "self" and "mutual" inductance.
            Give an example of the application of each of these effects.

     (b)   A conductor is moved at 10 m/s through a magnetic field having a flux
            density of 0.8 tesla.

            Calculate: (a)   the e.m.f. induced per meter length of the conductor,
                              and
                       (b)   the force per meter length opposing this movement if
                              the conductor is connected in a circuit having a total
                              resistance of 1.6 ohms.

     (Ans. (a) 8 V/m (b) 4 N/m)

26. What is meant by self-inductance?
     Define the unit in which it is measured.

     If a coil of 150 turns is linked with a flux of 0.01 Wb when carrying a current
     of 10 A, calculate the average induced e.m.f. if the current is uniformly
     reversed in 0.1 sec.

     (Ans. 30 V)

27. (a)   Explain what is meant by the term "self" and "mutual" inductance. Give
            an example of the application of each of these effects.

(b) A conductor is moved at 20 m/s through a magnetic field having a flux density of 0.8 Tesla.

Calculate: (a) the e.m.f. induced per metre length of the conductor, and

(b) the force per metre/length opposing this movement if the conductor is connected in a circuit having a total resistance of 2 ohms.

(Ans. (a) 16 V/m, (b) 6.4 N/m)

28. A centre-tapped coil of 500 turns is mounted on a magnetic core and has a self-inductance of 10 mH.

Calculate:- (i) the induced e.m.f. when the flux changes at the rate of 10 mWb/s;

(ii) the self-inductance of each half of the coil.

(iii) the energy stored when the coil carries a current of 5.0 A.

(Ans. (i) – 5 V (ii) 5 mH (iii) 0.125 J)

29. (a) (i) A current, varying in the manner shown in the diagram, is passed through a coil having an inductance of 200 mH. Calculate the e.m.f. induced in the coil during the periods *AB, BC* and *CD*.

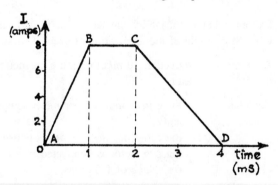

Fig. 12

(ii) Calculate the energy contained in the coil during the period *BC*.

(Ans. (i) – 1600 V, 0 V, 800 V (ii) 6.4 J)

(b) An ideal step-up transformer having a turns ratio of 1:5 is connected to a supply voltage of 100 V. The secondary load is a resistor whose resistance is 5 KΩ.

Calculate:- (i) the secondary voltage and current

(ii) the primary current

(iii) the effective resistance across the primary terminals.

232

30. (a) State Lenz's Law.

(b) A rectangular coil 30 cm long and 20 cm wide consisting of one turn is rotated about its longer axis at 700 rev/min in a uniform magnetic field of 0.5 tesla; the field being normal to the axis.

Calculate the maximum e.m.f. generated in the coil.

(c) The same coil is now placed at right angles to the field. At what rate must the field be changed in order that the statically induced e.m.f. shall be the same as the maximum e.m.f. generated when it was rotating.

(d) What torque would be required at this point if 1.5 A were passing through the coil?

(Ans. (b) $E = Blv$, $\omega r = v$, $E = 1.1$ V (c) 1.1 Wb/s (d) 0.045 Nm)

31. (a) State Lenz's Law.

(b) A conductor 30 cm in length and carrying a current of 100 mA is placed at right angles to a magnetic field having a flux density of 0.8 T.

Calculate the force on the conductor.

(c) At what speed must the conductor in (b) move at right angles to the same magnetic field in order that 2.4 V is induced in it?

(d) A 3:1 step-down transformer has a primary current and secondary voltage of 15 A and 100 V respectively. Calculate the primary voltage and secondary current.

(e) Calculate the energy contained in an inductance of 2 H carrying a current of 500 mA.

(Ans. (b) 0.024 N (c) 10 m/s (d) 33.3 V, 45 A (e) 0.25 J)

32. Explain what is meant by mutual inductance.

Give a practical example of where mutual inductance is essential and one where it is a disadvantage.

Two coils A and B are mutually coupled, the mutual inductance between them being 0.1 Henry.

If the coils have 100 turns each, determine the e.m.f. induced in coil B when

(i) the current in coil A changes at 50 A/s

(ii) the flux linking the coil changes at 0.3 weber/sec.

(Ans. (i) 5 V (ii) 30 V)

33. (a) A conductor at right angles to a uniform magnetic field of flux density 20 mT has 4 cm of its length in the field.

Calculate the current required to cause a force of 40 mN to act on the conductor.

(b) A current of 100 mA flowing in a 50 mH coil is reversed linearly in 200 μS.

Calculate the value of the e.m.f. induced in the coil, and sketch graphs of current and e.m.f. against time for this period.

Calculate also the energy stored in the magnetic field when the coil current is 100 mA.

34. (a) Show that $L \propto N^2$.

(b) The flux linked with a coil changes steadily from 10 mWb to 60 mWb in 10 ms. The average value of induced e.m.f. is 200 V. How many turns are there on the coil?

35. (a) (i) State Lenz's Law.

(ii) Calculate the e.m.f. induced in an inductance of 100 mH when the current is changing at a rate of 400 A/s.

(iii) Calculate the energy contained in this inductance when the current is 2A.

(b) (i) Explain the term "eddy current loss" when applied to transformers. How is this loss reduced?

(ii) A 4:1 step-down transformer has a primary voltage supply of 240 V and a secondary load resistance of 500. Calculate the resistance at the primary terminals.

(Ans. (a) (ii) 40 V (iii) 0.2 J (b) (ii) 8 KΩ)

---

36. (a) An alternating voltage is given by the expression

$$v = 50 \sin \left(3142t + \frac{\pi}{6} \right)$$

Calculate: (i) the peak, average and r.m.s. values;
(ii) the frequency, angular velocity and periodic time;
(iii) the instantaneous value of the voltage when t = 0.25 mS.

(b) Sketch the above waveform for a complete cycle

(Ans. (a) 50 V, 31.8 V, 35.3 V (b) 500 Hz, 3142 rad/s, 2 ms, 48.3 V)

37. An instantaneous current is given by the expression

$$i = 70.7 \sin \left(3124\,t + \frac{\pi}{3} \right) \text{ mA}$$

(a) Calculate: (i) its frequency
   (ii) its r.m.s. value
   (iii) its instantaneous value after 0.8 mS

(b) Calculate the power developed in a resistor of 400 Ω when alternating current above flows through it.

(c) Represent the above current and the instantaneous voltage $v = 50 \sin 3124\,t$ volts in a phasor diagram.

(Ans. (a) (i) 500 Hz (ii) 50 mA (iii) 35.34 mA (b) 1 W

38. Explain with reference to sinusoidal current the meaning of the terms (a) frequency (b) periodic time (c) amplitude.

Write down the mathematical expression representing a sinusoidal current having an r.m.s. value of 100 mA and frequency 1 KHz.

The r.m.s. voltage across an inductor of negligible resistance is 10 V when this current flows in it. Deduce the expression for this voltage.

Using the same axis sketch these current and voltage waveforms over one cycle showing clearly the phase relationship.

(Ans. $i = 100\sqrt{2} \times 10^{-3} \sin 2\pi\, 1000t$,

$$v = 10\sqrt{2} \sin \left(2\pi\, 1000t + \frac{\pi}{2} \right))$$

39. (a) What is meant by the term "r.m.s. value of an alternating current"?

(b) The voltages across three components in a circuit can be expressed as:-

$$v_1 = 25 \sin \omega t$$

$$v_2 = 40 \sin \left(\omega t + \frac{\pi}{3} \right)$$

$$v_3 = 20 \sin \left(\omega t - \frac{\pi}{2} \right).$$

Determine the resultant total voltage across the circuit and express it in the form $v_R = V_m \sin (\omega \mp \phi)$

$\left(\text{Ans. } v_R = 47.5 \sin (\omega t + 0.332^c) \right)$

40. (a) Sketch one cycle of a 100 V 50 Hz sinusoidally varying voltage. Label the diagram and indicate the value of the periodic time and the amplitude.

(b) State the r.m.s. and average values of this voltage.

(c) Calculate the power that would be dissipated if this voltage were applied to a 20 ohm resistor.

(d) Another 100 V 50 Hz sinusoidally varying voltage is to be added to the voltage in 40 (a). They are displaced from each other by 90°.

Draw a phasor diagram representing these voltages and their resultant. State their resultant in the form

$$v = V_m \sin (\omega t + \phi).$$

(Ans.   (a) 20 ms, 141.4 V   (b) 100 V, 141.4 V   (c) 500 W

$$\text{(d) } v = 200 \sin \left[2\pi 50t + \frac{\pi}{4} \right] \Big]$$

41. (a) Explain with reference to sinusoidal current the meaning of the following terms:- (i)   periodic time

(ii)   frequency

(iii)   amplitude.

(b) Draw phasors to represent the following currents

$$i_1 = 6 \sin \omega t \qquad i_2 = 10 \sin \left[\omega t + \frac{\pi}{4} \right].$$

Determine the resultant current $i_1 + i_2$.

(c) Is the above resultant current inductive or capacitive?

(d) From your answer in (c), determine the following:-

(i)   the peak value of $i_1 + i_2$

(ii)   the frequency of $\omega = 2500$ radians.

(Ans.   (a) ---   (b) $i = 15.4 \sin \left[\omega t + \frac{\pi}{6} \right]$ (c) leading by $\frac{\pi}{6}$, capacitive

(d) (i) 15.4 A   (ii) 398 Hz.)

42. A supply voltage of $v = 141.4 \sin 3142t$ is applied to a circuit containing 25 Ω resistance and 20 mH inductance.

(a) State the frequency of supply voltage.

236

(b) Calculate (i) the current that would flow. State it in the form

$$i = I_m \sin (\omega t + \phi)$$

(ii) the power dissipated

(iii) the power factor.

Explain with the aid of suitable diagrams, the basic construction and operation of EITHER ONE of the following:-

(i) a permanent magnet moving coil instrument

(ii) a C.R.O.

(iii) a Wheatstone Bridge.

(Ans. (a) 500 Hz   (b)   (i) 1.48 A, $i = 2.03 \sin (2\pi 500t - 1.19^c)$

(ii) 155 W   (iii) 0.623).

43. A sinusoidal alternating voltage has the equation

$$v = 141.4 \sin 377 \, t.$$

Determine (a) the peak, average and r.m.s. values,

(b) the angular velocity, frequency and periodic time,

(c) the instantaneous voltage when t = 3 ms.

(Ans. (a) 141.4 V, 90 V, 100 V (b) 377 rad/s, 60 Hz, 16.7 ms (c) 127.9 V.)

44. A series circuit consists of a coil of inductance 0.3 H and resistance 25 Ω and a capacitor.

When a voltage of 240 V is applied the circuit resonates at a frequency of 50 Hz.

Calculate (i) the value of C;
(ii) the current at resonance;
(iii) the voltage across the inductor;
(iv) the power dissipated;
(v) the Q-factor of the circuit

(Ans. (i) 33.8 μF (ii) 9.6 A (iii) 905 V (iv) 2304 W (v) 3.8)

45. (a) Sketch neatly a sine wave of current, and in relation to this current sketch the voltage waveforms you would expect when this current passes through (i) a resistance,
(ii) an inductance,
(iii) a capacitance.

Label your diagram.

237

(b) In the circuit shown in Fig. 13 calculate

    (i)   the supply current

    (ii)   the supply phase angle

    (iii) the supply power.

$$f = 125 \text{ Hz}$$

Fig. 13

(Ans. (a) --- (b) (i) 0.2 A (ii) 53° 8′ (iii) 12 W)

46. A coil takes a current of 2 A when connected to a 100 V d.c. supply. When the same coil is connected to a 100 V 50 Hz a.c. supply, the current taken is 0.5 A.

Explain why these currents differ.

Calculate (a)   the resistance of the coil,

            (b)   the impedance of coil at 50 Hz

            (c)   the power taken from the d.c. supply.

(Ans. (a) 50 Ω (b) 200 Ω (c) 200 W).

47. For the circuit shown, calculate:-

    (a)   the total current;

    (b)   the power factor

    (c)   the voltage across each component.

$$f = 1 \text{ kHz}$$

Fig. 14

(Ans. (a) 188 mA (b) 0.94 (c) 1.88 V, 1.77 V, 2.44 V)

8. (a) In the circuit shown below, calculate the following:-
  (i) the total impedance and phase angle of the circuit,
  (ii) the voltage across the capacitor,
  (iii) the power developed in the circuit.

f = 1500 Hz
Fig. 15

(b) Describe briefly what must be done to the circuit to improve the power factor to unity.

(Ans. (a) (i) 263 Ω, 67° 38′ (ii) 10.08 V (iii) 0.9 W)

9. (a) In the circuit shown in Fig. 16 calculate:-
  (i) the supply current
  (ii) the supply power

(b) What value of capacitance must be added to the circuit in order that series resonance should occur?

  In this case what will be (i) the supply current,
                             (ii) the supply power.

f = 1 kHz
Fig. 16

(Ans. (a) (i) 0.1 A (ii) 4 W (b) 0.53 μF (i) $\frac{1}{8}$ A (ii) 6.25 W)

50. A capacitance of 0.10 μF and an inductor of resistance 40 Ω and inductance 2.0 mH are in series across a supply of 5.0 V at the resonant frequency.

Calculate the capacitor voltage and the power dissipated.

Draw a phasor diagram to show all the voltages and the current.

(Ans. 17.7 V, 0.625 W)

51. (a) A circuit breaker coil for use on 240 V 50 Hz is tested by an ohmmeter and gives a reading of 300 Ω. When operating on the coil takes 0.6 amperes.

Find (i)   the coil impedance;

(ii)   the coil inductance.

(b) Calculate the coil reactance at a frequency of 25 Hz.

(Ans. (a) (i) 400 Ω (ii) 0.842 H (b) 132 Ω)

52. A series circuit consists of a non-inductive resistance of 10 Ω, a capacitance of 20 μF, and a pure inductance of 100 mH. The supply voltage is 100 V r.m.s. at 150 Hz.

Calculate:-   (i)   the circuit impedance,

(ii)   the current,

(iii) the phase angle.

Draw a phasor diagram for the above circuit.

(Ans. (i) 42.3 Ω (ii) 2.36 A (iii) 76.3°)

53. A circuit having an inductance of 15.9 mH and a resistance of 8 Ω in series carries a current of 24 A at a frequency of 50 Hz.

Determine:-   (a)   the impedance of the circuit

(b)   the applied voltage

(c)   the power dissipated

(d)   the volt-amperes supplied

(e)   the power factor.

(Ans. (a) 9.43 Ω (b) 226 V (c)4608 W (d) 5424 VA (e) 0.85)

54. A coil of resistance 15 Ω and inductance 63.66 mH is connected across a 200 V, 50 Hz sinusoidal supply. Calculate the current, phase angle and power dissipation. Sketch, approximately to scale, the phasor diagram.

(Ans. 8 A, 53.13°, 960 W)

55. (a) Explain with the aid of a circuit diagram, how a simple Wheatstone bridge may be used to measure the value of an unknown resistor.

State using this circuit diagram, the conditions necessary for balance.

What main precaution should be taken when using this bridge?

(b)   Compare the Wheatstone bridge with the ammeter and voltmeter methods for measuring resistance.

56.   An ammeter and voltmeter are to be used to obtain the value of resistance of a circuit. Sketch both possible methods of connection and explain why, in each case, it would be incorrect to assume that the resistance is given by the voltmeter reading divided by the ammeter reading.

57.   Describe, with the aid of a simple diagram, the construction and operation of ONE of the following:-   (i)   a moving coil instrument,

(ii)   a cathode ray oscilloscope

(iii)   a Wheatstone bridge.

58.   Explain with a circuit diagram what is meant by a potential divider, and show how two potential dividers can make a Wheatstone Bridge circuit.

Write the equation which is used to calculate an unknown resistance when using a Wheatstone Bridge, showing which resistance is which in a circuit diagram.

Explain why this is not a good method for measuring very high resistance.

59.   Describe, with the aid of a simple diagram, the construction and operation of ONE of the following:-

(i)   a moving coil instrument,

(ii)   a cathode ray oscilloscope

(iii)   a Wheatstone bridge.

60.   (a)   A 0 to 10 V meter has a resistance of 100000 Ω. What additional voltage multiplier resistor is required to enable the meter to indicate 0 to 250 V.

(b)   For the Wheatstone balanced bridge shown below, determine

(i)   the value of R_x

(ii)   the p.d. across R_x

(iii)   the current through DC.

(Ans. (a) 2.4 MΩ (b) (i) 10 Ω (ii) 1 V (iii) 5.88 mA)

241

Fig. 17

61. (a) A moving coil meter movement has a coil resistance of 350 Ω and resistor of 650 Ω is connected in series with it. The movement require a current of 1 mA for f.s.d.

    Calculate the shunt resistance required to change the movement into 0-500 mA ammeter.

    (b) Explain, with the aid of a suitable diagram, the principle of operation the Wheatstone bridge. Derive from first principles an expressio giving, for the balance condition, the unknown resistance in terms of th known quantities.

    (Ans. (a) 2.004 Ω (b) $R_x = \left(\dfrac{R_1}{R_2}\right) R_3$)

62. A moving coil milliammeter has a coil resistance of 5 Ω and gives full scal deflection for a current of 15 mA. The scale has 15 divisions.

    (i) Calculate the value of the resistor required to enable the instrument t read 0-3 A. By what factor must the scale reading be multiplied to giv correct readings of current?

    (ii) Calculate the value of the resistor required to enable the milliammete to read 0-150 V. How many volts will each division represent?

    (Ans. (i) 0.0251256 Ω, 20 (ii) 9995 Ω, 10 V)

63. Sketch the basic construction of a moving-coil instrument, and explain th principle of operation. How is damping achieved?

242

64. A sinusoidal p.d. $v = 20 \sin 3142t$ is applied to a 15.92 μF capacitor.

Calculate:-   (i)   the frequency,
            (ii)   the r.m.s. value of the voltage
            (iii)   the reactance of the capacitor
            (iv)   the r.m.s. value of current.

Write the equation for the current, and sketch both the wave diagram and the phasor diagram.

(Ans.   (i) 500 Hz (ii) 14.14 V (iii) 20 Ω (iv) 0.707 A

$$i = \sin \left(3142 + \frac{\pi}{2} \right))$$

65. (a)   The alternating current in a circuit of resistance 80 Ω is represented by
$$i = 3 \sin 100 \, \pi t$$
where i is in ampere and t in seconds.

Sketch the waveforms of current and voltage for the first 20 ms.

Write down   (i)   the peak value of this current
             (ii)   its r.m.s. value
             (iii)   its frequency
             (iv)   its period

(b)   If the resistor is replaced by a capacitor of 40 μF and the same value of current flows. Give an expression for the voltage across it.

(Ans.   (a) 1 cycle (i) 3 A (ii) 2.121 A (iii) 50 Hz (iv) 20 ms

$$(b) \; v = 240 \sin \left(100 \pi t - \frac{\pi}{2} \right)$$

66. An instantaneous current in a circuit is given by the expression

$$i = 100 \sin \left(31420t + \frac{\pi}{4} \right) \text{ mA.}$$

(a)   Calculate   (i)   the frequency
               (ii)   its r.m.s. value
               (iii)   its instantaneous value when $t = 25 \; \mu$s.

(b)   Calculate the power developed in a 1 kΩ resistor when the above current flows through it.

(c)   If the instantaneous value of voltage supplied to the circuit is $v = 50 \sin 31420t$ volts, draw a phasor diagram showing the r.m.s. values of voltage and current.

(Ans. (a) (i) 5000 Hz (ii) 70.7 mA (iii) 100 mA (b) 5 W)

67. (a) An alternating voltage is represented by the expression

$$v_1 = 20 \sin 1257t \text{ volts}$$

Determine:- (i) the frequency of the wave,
(ii) the periodic time,
(iii) the times taken from $t = 0$ for the voltage to become $+ 15$ V for the first time in the cycle.

(b) if a second alternating voltage is represented by the expression

$$v_2 = 15 \cos 1257t,$$

add the two voltages and express the resultant in the same form.

(Ans. (a) (i) 200 Hz (ii) 5 ms (iii) 675 μs (b) $v = 25 \sin (1257t - 0.644°)$)

68.

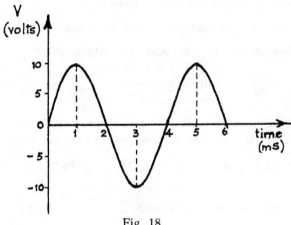

Fig. 18

For the given sinusoidal voltage state the values of

(i) amplitude
(ii) period
(iii) frequency
(iv) r.m.s. voltage
(v) peak to peak voltage
(vi) instantaneous voltage at time 1 ms.

(Ans. (i) 10 V (ii) 4 ms (iii) 250 Hz (iv) 7.07 V (v) 20 V (vi) 7.07 V)

PHASE TEST. 1 (69 - 80 inclusive)

69. For the figure 19,

Fig. 19

write down the magnetomotive force and the units in which it is measured.

70. What is the ratio $\dfrac{B}{H}$ equal to ?

71. Current flows

Fig. 20

as shown, mark the poles of the electromagnet.

72. Sketch the magnetization curve of a ferromagnetic material, label the axes and the units in which the physical quantities are measured in.

73. Sketch the hysteresis loop of a soft material labelling the axes and insert the units in which the physical quantities are measured in.

74. Write down the total reluctance of a magnetic circuit which is made up of two circuits having the same flux but different permeabilities $\left(\mu_{r_1} , \ \mu_{r_2} \right)$.

75. Give the unit of reluctance.

76. Sketch the magnetic field of a permanent magnet.

77. The magnetizing force in a magnetic circuit is 5000 At/m. Find the number of turns if a current is flowing of 5 A and the length of the magnetic path is 0.5 m.

78. Define the term of relative permeability.

79. The self inductance of a coil in 5 H, and the reluctance is 1,000,000 A/Wb, calculate the number of turns.

80. Show that the input a.c. resistance viewed at the primary of the transformer with a load of R_L is given by the formula $R_{in} = \left(\dfrac{N_p}{N_s}\right)^2 R_L$.

81. A current, varying in the manner shown in the diagram, is passed through a coil having a self inductance of 0.5 H.

Calculate the e.m.f. induced in the coil during the periods   (a)   *OA*
                                                              (b)   *AB*.

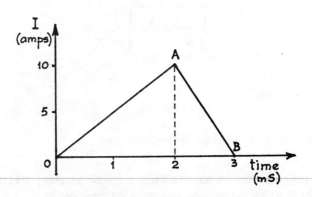

Fig. 21

82. The energy stored in a coil is 50 mJ and the self-inductance is 50 mH. Determine the current flowing.

83. If the mutual inductance of the coils is 0.05 H determine the total self inductance between the terminals *A* and *B*. Insert the dots to indicate the direction of the fluxes set up.

246

Fig. 22

. The rate of change of current in N_1 of the primary of a transformer when the switch is closed is 25 A/s, the mutual inductance is 1 H calculate the induced e.m.f. across R. If the rate of flux is 500 mWb/s linking the coils, determine the number of turns, N_2 of the secondary.

PHASE TEST. 2 (85 - 88 inclusive)

. (i) Calculate the capacitance C_{AB}

Fig. 23

(ii) Calculate the capacitance C_{CD}

Fig. 24

(iii) A d.c. voltage is applied across CD giving rise to 100 μC of a charge on the 25 μF.

247

Calculate (a) the d.c. voltage applied across *CD*

(b) the energy stored in one of the 50 μF capacitor.

86.

Fig. 25

Determine the current I_1 and I_2 using Kirchhoff's Laws and hence the load current, indicating its direction.

87. A coil of 1000 turns is wound round a toroidal magnetic material of relative permeability, $\mu_r = 500$. The cross sectional area $A = 100$ mm^2 and the mean magnetic length is 35 π cm. Calculate the magnetizing force if the d.c current flowing through the coil is 5 A.

88.

Fig. 26

Determine (i) V_1

(ii) V_2.

89.

Fig. 27

Determine I_1 and I_2.

90.

Fig. 28

Give the formula for the total capacitance and prove it.

91.

Fig. 29

Calculate I_1 using the current divider and prove the formula used.

92.

Fig. 30

Determine H the magnetizing force if the mean magnetic length is 150 mm and the reluctance of the circuit.

93.

Fig. 31

Calculate   (i)   the total charge Q,

               (ii)   the energy stored in the system.

PHASE TEST. 4 (94 - 98 inclusive)

94.

Fig. 32

Determine the currents I_1 and I_2 by using the current divider rule. Prove the formula used.

Fig. 33

Determine V_1 by using the potential divider rule. Prove the formula used.

Fig. 34

Determine the currents I_1 and I_2 using Kirchhoff's second law.

Fig. 35

251

Calculate (a) the charge on each capacitor before A is connected

(b) the total energy stored and

(c) the value of A which when connected in parallel with the 10 μF capacitor would change V_2 to 80 V with the same supply.

98. A ring specimen of cast steel has a radial air gap 1 mm long cut through th ring. The ring has a mean length of 40 cm and c.s.a. of 8 cm^2. If 200 turn of wire are would uniformly round the ring, calculate the current required t set up a magnetic flux of 640 μWb in the ring and gap.

$\mu_r = 800$, $\mu_o = 4\pi\ 10^{-7}$ H/m.

PHASE TEST. 5 (99 - 101 inclusive)

99. A capacitor is charged with 8 mC. If the energy stored is 0.4 J calculate:-

(a) the voltage,

(b) the capacitance.

100. A capacitor, consisting of two metal plates each of area 5000 mm^2 and spaced 0.2 mm apart in air, is connected across a 120 V supply. Calculate:-

(a) the energy stored,

(b) the electric flux density,

(c) the potential gradient.

101. Three capacitors have capacitances of 1 μF, 2 μF, and 3 μF respectively Determine the total capacitance of the three capacitors when they are connected (i) in parallel

(ii) in series.

A capacitance of 2.2 μF is required.

Connect the above capacitors in a suitable circuit to achieve this value, and calculate the total energy stored if a p.d. of 100 V d.c. is applied across the whole circuit.

PHASE TEST. 6 (102 - 103 inclusive)

102. Spell out the units for the following:-

(i) Flux

(ii) Flux density

(iii) Magnetic field strength

(iv) Permeability (absolute)

(v) Magnetomotive force

(vi) Reluctance

(vii) Relative permeability

(viii) Electromotive force

(ix) The area of the hysteresis loop

(x) Coersive force.

03. A coil of 500 turns wound on ring of magnetic material, carries a current of 1 A. The length and cross-sectional area of the flux path in the ring are 500 mm and 1000 mm^2 respectively. The relative permeability μ_r = 3500 and μ_o = $4\pi \times 10^{-7}$ H/m.

Determine (i) the magnetic flux density (ii) the reluctance.

04. (a) Make a neat sketch of a hysteresis loop and with its and explain the terms:-

 (i) remanent flux density, B_r,

 (ii) coersive force, H_c

 (iii) initial magnetisation

 (iv) magnetic saturation.

 (b) A ring specimen of cast steel has radial air gap 0.5 mm long cut through the ring and the air gap has a reluctance of 5×10^5 A/Wb. The ring has a mean length of 50 cm and relative permeability of 1000. Calculate the number of turns wound on the ring if 1.5 A is required in these turns to set up a flux of 500 mWb.

PHASE TEST. 7 (105 - 108 inclusive)

05. A conductor 50 cm in length and carrying a current of 0.5 A is placed at 80° to a magnetic field having a flux density of 1.5 T. Calculate the force on the conductor.

06. A conductor is moved at 15 m/s through a magnetic field having a flux density of 2 T.

Calculate (a) the e.m.f. induced per meter length of the conductor

 (b) the force per meter length opposing this movement if the conductor is connected in a circuit having a total resistance of 1.5 Ω.

107. A coil of 1000 turns is subjected to a flux change at the rate of 20 mWb/s. Determine the magnitude of the induced e.m.f. in the coil.

108. The flux linked with a coil changes steadily from 5 mWb to 80 mWb in 8 ms. The average value of induced e.m.f. is 100 V. How many turns are there on the coil.

(i) Calculate the e.m.f. induced in an inductance of 500 mH when the current is changing at a rate of 200 A/s.

(ii) Calculate the energy stored in this inductance when the current is 5 A.

PHASE TEST. 8 (109 - 111 inclusive)

109. Determine the currents I_1, I_2, I_3 and I_4 in the circuit of Fig. 36, showing clearly all the equations and calculations.

Fig. 36

Calculate the total power dissipated in the circuit and the energy consumed in joules, in 1 hour.

110. Determine the currents I_1 and I_2 in Fig. 37, using Kirchhoff's Laws and hence find the load current, indicating its direction.

Fig. 37

254

111. (i)   Calculate the capacitance C_{AB}

Fig. 38

(ii)   Calculate the capacitance C_{CD}

$C_5 = C_4 = 50\ \mu F$

$C_6 = 25\ \mu F$

Fig. 39

(iii)   A d.c. voltage is applied across CD giving rise to 100 μC of a charge on the 25 μF.

Calculate:-

        (a)   the d.c. voltage applied across CD

        (b)   the energy stored in one of the 50 μF capacitors.

PHASE TEST. 9 (112 - 115 inclusive)

112.

Fig. 40

Determine the value of the resistance R.

113

Fig. 41

Determine V_1 and V_2.

114.

Fig. 42

Calculate I and hence R_T, check using the formula above. Calculate also the power dissipated in each resistor.

115.

Fig. 43

Find I_1 and I_2.

PHASE TEST. 10 (116 - 120 inclusive)

116.

Fig. 44

Determine the branch currents.

117.

Fig. 45

Use the current divider to work out the current I_1 and I_2

118.

Fig. 46

Assume I flows as shown. Mark the p.d.s across the active and passive components, and use the K.V.L.to find I. Emphasis on the Kirchhoff's is required.

119.

Fig. 47

Show that:- $I_1 = \dfrac{R_2}{R_1 + R_2} I$

$I_2 = \dfrac{R_1}{R_1 + R_2} I$

258

20.

Fig. 48

Find I_1 and R.

PHASE TEST. 11 (121 - 122 inclusive)

21. (a)   An alternating voltage is represented by the expression

$$v_1 = 20 \sin 1257t \text{ volts}$$

Determine:-   (i)   the frequency of the wave,
              (ii)   the periodic time,
              (iii) the time taken from $t = 0$ for the voltage to become
                     $+ 15$ V for the first time in the cycle.

(b)   If a second alternating voltage is represented by the expression

$$v_2 = 15 \cos 1257t,$$

add the two voltages and express the resultant in the same form.

22.

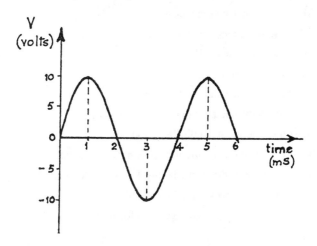

Fig. 49

259

For the given sinusoidal voltage state the values of:-

(i)   amplitude

(ii)   period

(iii)   frequency

(iv)   r.m.s. voltage

(v)   peak to peak voltage

(vi)   instantaneous voltage at time $\frac{1}{2}$ ms.

PHASE TEST. 12 (123)

123.

Fig. 50

Calculate the following in 3 S.F.

(i)   the inductive reactance

(ii)   the impedance of the circuit

(iv)   the p.d. across R

(v)   the p.d. across L

(vi)   the r.m.s. value of the current

(vii)   the impedance triangle, label the sides

(viii) the voltage triangle, label the sides

(ix)   the formula for the a.c. power and calculate it

(x)   the phasor diagram, taking I as the reference, label clearly

(xi)   the power factor of the circuit

(xii)   the current and voltage waveforms (not to scale)

24.

Fig. 51

Determine the value of the resistor R, if R_1 = 25 Ω.

25.

Fig. 52

Determine the values of V_1 and V_2

26.

Fig. 53

Calculate I and hence R_T, check your answer using the formula above. Calculate also the power dissipated in each resistor.

127.

Fig. 54

Determine the currents I_1 and I_2.

PHASE TEST. 14 (128 - 131 inclusive)

128.

Fig. 55

Determine the load current I_L using Kirchhoff's laws.

129.

$R_1 = 50\ \Omega$

$R_2 = 100\ \Omega$

Fig. 56

(a) Determine V_o.

(b) Determine V_o, if another resistor $R_3 = 100\ \Omega$ is connected across the combination R_1, R_2.

130.

Fig. 57

Determine (a) I (b) I_1 and I_2

131.

$S = 500,000\ A/Wb$

$\mu_o = 4\pi \times 10^{-7}\ H/m$

Fig. 58

Determine (a) the magnetic field strength
       (b) the magnetic flux
       (c) the magnetic flux density
       (d) the m.m.f.
       (e) μ_r.

END TEST. 1 (132 - 137 inclusive)

132. (a) Use the potential divider method to determine the p.d.s V_1 and V_2 shown in the diagram.

Fig. 59

   (b) Determine the currents I_1, I_2 and I_3

133. Determine the currents I_1, I_2 and I_L in the circuit using Kirchhoff's Laws.

Fig. 60

134. Show that two capacitors when they are connected in series the total value is

   given $C = \dfrac{C_1 C_2}{C_1 + C_2}$. Two capacitors 10 μF and 30 μF are connected in

   series. Across both of these is connected a capacitor of 20 μF. The whole combination is connected across a 100 V d.c. supply.

Calculate:- (i)   the equivalent capacitance
            (ii) the p.d. across the 30 μF capacitor
            (iii) the charge on 10 μF capacitor.

35. A piece of iron has a magnetic length of 150 mm a c.s.a. of 500 mm², and a relative permeability of 550. Find the reluctance of iron.

$\left(\mu_o = 4\pi 10^{-7} \text{ H/m}\right)$. Determine the length of the radial cut required such that the reluctance is the same as that of the iron. Calculate the total m.m.f. required if the flux set up in both the ring and the air gap is 50 μWb.

36. (a) An instantaneous voltage is given by the expression

$$v = 100 \sin\left(3142t + \frac{\pi}{6}\right).$$

Calculate:- (i)   the frequency
            (ii) the r.m.s. value
            (iii) the instantaneous value after 1 ms.

(b)

Fig. 61

Calculate:- (i)   the total impedance
            (ii) the power developed in the circuit
            (iii) the r.m.s. current.

37. Construct a simple ohm-meter using an ammeter with a full-scale deflection of 15 mA a variable resistor, and a 15 V battery. Draw the circuit for the arrangement.

Calculate the resistance indicated by the instrument when currents of

(i)   1 mA
(ii) 5 mA and
(iii) 10 mA are flowing in the circuits.

138.

Fig. 62

Calculate:- (a) the total resistance of the network.

(b) the total current, I.

(c) the p.d. across A and B

(d) the p.d. across C and D

(e) the power in the 50 Ω resistor.

139.

Fig. 63

(a) State clearly the Kirchhoff's current and voltage laws, illustrating them using the adjacent diagram.

(b) Calculate the load current through the 10 Ω resistor.

140.

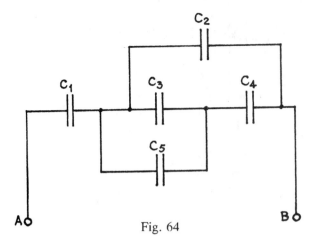

Fig. 64

(a) Explain clearly the factors on which the capacitance of a capacitor depend. Write down the relative formula for 10 plates.

(b) The thicknesses of the dielectrics of the capacitors C_1, C_2, C_3, C_4 and C_5 are 1mm, 2mm, 3mm, 4mm, and 5mm respectively. Assuming that each dielectric has a relative permittivity of 10, $\in_o$ = 8.85 × 10^{-12} F/m and there are two parallel plates in each capacitor of cross sectional area of 10 cm^2.

Calculate the total value of the capacitance between A and B.

If V_{AB} = 200 volts, calculate the total energy.

141. (a) A ring specimen made of a steel-alloy has a mean diameter of 50 cm and a cross sectional area of 10 cm^2. In a certain magnetic test the reluctance of the ring is found to be 6 × 10^6 A/wb. Calculate the relative permeability of the steel-alloy in this case

$\left(\mu_o = 4 \times 10^{-7} \text{ H/m}\right)$.

(b) A radial air gap is now made in this ring and it is uniformly wound with a coil of 500 turns. When a current of 1 A flows in the coil, it is found that the reluctance of the air gap is the same as that of the ring. Determine the magnetic flux set up in the ring.

142. (a) (i) A current varying in the manner shown in the diagram is passed through a coil having an inductance of 0.5 H.

Calculate the e.m.f. induced in the coil during the periods AB, BC and CD.

(ii) Calculate the energy contained in the coil during the period BC.

267

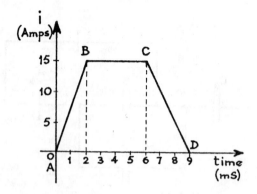

Fig. 65

(b) An ideal step-down transformer having a turns ratio 20:1 is connected to a supply of 240 V. The secondary load is a resistor whose resistance is 1000 Ω.

Calculate:- (i)   the secondary current and voltage

(ii) the primary current

(iii) the resistance viewed at the input terminals.

143. (a) Explain with the aid of suitable diagrams the basic construction and operation of <u>one</u> of the following:-

(i)   a Wheatstone Bridge

(ii) a permanent magnet moving coil instrument

(iii) an a.c. rectifier meter.

(b) A supply voltage of $v = 141.4 \sin (3142t + 30°)$ is applied to a circuit containing 50 Ω resistance and 25 mH inductance.

(i)   State the frequency of the supply voltage

(ii) Calculate the current that would flow in the form

$$i = I_m \sin \left(\omega t \pm \phi\right)$$

(iii) the power dissipated

(iv) the power factor.

144. A circuit having an inductance of 20 mH and a resistance of 5 Ω in series carries a current of 25 A at a frequency of 50 Hz.

Determine:- (a) the impedance of the circuit

(b) the applied voltage

(c) the power dissipated

(d) the volt-amperes supplied

(e) the power factor.

END TEST. 3 (145 - 151 inclusive)

145. (a)   Using Kirchhoff's Law in Fig. 66, find the current in each branch.

    (b)   Find the power dissipated in resistor R_3 in Fig. 66.

Fig. 66

146. In the circuit shown in Fig. 67 find:-

    (a)   The equivalent capacitance of the circuit.

    (b)   The voltage across C_1.

    (c)   The charge on each capacitor.

    (d)   The energy stored in each capacitor.

Fig. 67

147. (a) State the direction of the force acting on the above current carrying conductor.

Fig. 68

(b) One side of a square coil of side length 5 cm is a right angle to a magnetic field of flux density 50 mT. When the coil current is 100 mA, the force acting on the coil side is 20 mN. Calculate the number of turns on the coil.

(c) The coil of a 60 mH inductor has a resistance of 200 Ω. Calculate the energy stored by the magnetic field when the coil is connected across a 50 V supply.

148. (a) Sketch the hysteresis loop for a soft magnetic material, naming the axes.

(b) A cast iron toroid of mean circumference 15 cm and cross-sectional area 1 cm² is wound with 500 turns. A 0.2 mm radial air gap is cut in the toroid. Calculate the current required to establish a flux of 600 μWb in the toroid. Neglect leakage and fringing flux. Points on the characteristic of cast iron are shown below.

B mT	310	500	600	640
H A/m	1000	2000	3000	4000

$\mu_o = 4\pi \times 10^{-7}$ H/m.

149.

Fig. 69

270

For the above waveform, determine:-

    (a)   The average and r.m.s. values.

    (b)   The frequency.

    (c)   The form factor.

50. A 12 KΩ resistor is connected in series with a 994.7 pF capacitor across a 200 mV, 10 kHz supply.

    Calculate:-   (a)   The current and phase angle.

                   (b)   The p.d. across each component.

                   (c)   The power dissipation.

    Sketch, approximately to scale, the phasor diagram.

51. A moving-coil instrument of internal resistance 5 Ω has full scale deflection at 5 mA. Calculate the values of resistors required to give the following additional ranges:- (a)   25 mA.

                                          (b)   10 V.

    Sketch the circuit diagram, give all resistance values and switches necessary for all 3 ranges. Show as a table the switch positions required for each range.

END TEST. 4 (152 - 158 inclusive)

52. Show that $I_1 = \dfrac{R_2}{R_1 + R_2} I$

                  $I_2 = \dfrac{R_1}{R_1 + R_2} I$

Fig. 70

271

(b)   Show that $V_1 = \dfrac{R_1}{R_1 + R_2} V$

Fig. 71

(c)   Determine the currents I_1, I_2 and I_3.

Fig. 72

153. Use Kirchhoff's Laws to obtain the values of the currents I_1, I_2 and I_3.

Fig. 73

272

54. (a) Show that the capacitance of a parallel plate capacitor is given by the formula.

$$C = \frac{\epsilon_o \epsilon_r A}{d}$$

where ϵ_o, ϵ_r are the permittivities of free space and relative respectively, A is the cross sectional area, and d is the separation of the plates.

(b) A capacitor is formed by two parallel metal plates each of area 1,000 mm² and separated by a dielectric 0.5 mm thick. The capacitance of this arrangement is 500 pF and a d.c. p.d. of 20,000 V is applied to the terminals.

Calculate:- (i) the charge on the plates
(ii) the relative permittivity of the dielectric
(iii) the electric flux density
(iv) the energy stored.

$(\epsilon_o = 8.85 \times 10^{-17}$ F/m)

55. (a) Draw the magnetic field around a permanent magnet.

(b) Draw the circuit of an electro-magnet showing the direction of the flux.

(c) A coil of 500 turns, wound on a ring of magnetic material carries a current of 1 A. The length and cross-sectional area of the flux path in the ring are 500 mm and 1000 mm² respectively.

The relative permeability = 3500 and the permeability of free space = $4\pi \times 10^{-7}$ H/m.

Determine:- (i) the magnetic flux density
(ii) the reluctance.

56. (a) By means of clear diagrams illustrate the motor principle (lefthand rule).

(b) State the **two** Faraday's Laws illustrating the induced e.m.f. by considering the rates of change of flux and currents, stating the unit for each.

By means of diagrams, explain clearly Lenz's Law.

(c) Using Faraday's Law, establish the formula for the induced e.m.f. in a conductor cutting the field at right angles with a velocity, $(E = Blv)$.

57. An instantaneous voltage is given by the expression $= 100 \sin \left(3143t + \frac{\pi}{6} \right)$

273

(a) Calculate:- (i) the frequency
(ii) the period
(iii) the r.m.s. value
(iv) the average value
(v) the form factor
(vi) the instantaneous value after 1 ms.

(b) Represent the above voltage and the instantaneous current $i = 100 \sin 3142\, t$ (a) in a phasor diagram.

158. (a) An ammeter of resistance 0.05 Ω has a shunt of resistance 0.015 : connected across it. What is the total current flowing when the curre thorough the ammeter is 5 A?

(b) In the circuit shown, calculate:-

Fig. 74

(i) the total impedance and phase angle of the circuit
(ii) the voltage across the coil
(iii) the power factor of the circuit.

END TEST. 5 (159 - 163 inclusive)

159. (a) Referring to figure 74 below, write equations which show the meanin of Kirchhoff's laws applied to:-

(i) junction B, and
(ii) loop $CBEDE$.

Fig. 75

274

(b)   A capacitor is formed by using two parallel plates, each of area 5000 mm^2 and separated by dielectric of 0.5 mm thick and relative permittivity of 50. A p.d. of 10 KV is applied to the terminals.

Calculate:-   (i)   the capacitance of the capacitor;

(ii)   the charge on the plates; and

(iii)   the energy stored.

160. A coil of 100 turns is wound uniformly on a non-magnetic material. The ring has a mean circumference of 0.5 metre and a uniform across sectional area of 0.003 square metre. If the current flowing through the coil is 2 amperes calculate:- (a)   the magnetomotive force,

(b)   the magnetising force

(c)   the flux density

(d)   the total flux

(e)   the total reluctance.

161. A current of 5 A in a coil of 900 turns sets up a flux of 500 μWb in the windings. Assuming this flux links all the turns.

Calculate:-   (a)   the self-inductance of the coil,

(b)   the energy stored in the coil when carrying a current of 5 A,

(c)   the e.m.f. induced in the coil if the current of 5 A is reversed in 50 msec.

162. An alternating voltage is represented by the equation $V = 150 \sin 3140t$ volts.

Determine the following:-   (a)   amplitude,

(b)   frequency,

(c)   periodic time

(d)   the r.m.s. voltage

(e)   peak-to-peak voltage and

(f)   instantaneous value when $t = 0.5$ msec.

163. A single-phase, a.c. circuit consists of a coil of resistance 80 ohms and inductance of 200 mH in series with 10 μF capacitor. The supply voltage is 240 V at 200 Hz.

Calculate:-   (a)   the reactance of the coil,

(b)   the reactance of the capacitor,

(c)   the circuit impedance,

(d)   the current and

(e)   the phase angle, stating whether lagging or leading.

275

END TEST. 6 (164 - 168 inclusive)

164. (a) Use the potential divider method to determine the p.d.s V_1 and V_2 shown in the diagram.

Fig. 76

(b) Determine the currents I_1, I_2 and I_3.

Fig. 77

165. Determine the currents I_1, I_2 and I_L in the circuit using Kirchhoff's Laws.

Fig. 78

276

66. Show that two capacitors when they are connected in series the total

value is given $$C = \frac{C_1 C_2}{C_1 + C_2}.$$

Two capacitors 10 μF and 30 μF are connected in series. Across both of these is connected a capacitor of 20 μF. The whole combination is connected across a 100 V d.c. supply.

Calculate:- (i)   the equivalent capacitance
(ii) the p.d. across the 30 μF capacitor
(iii) the charge on 10 μF capacitor.

67. A piece of iron has a magnetic length of 150 mm a c.s.a. of 500 mm^2, and a relative permeability of 500. Find the reluctance of the iron. ($\mu = 4\pi 10^{-7}$ H/m).

Determine the length of the radial cut required such that the reluctance is the same as that of iron. Calculate the total m.m.f. required if the flux set up in both the ring and the air gap is 50 mWb.

68. (a)   An instantaneous voltage is given by the expression

$$v = 100 \sin \left(3142t + \frac{\pi}{6} \right).$$

Calculate:- (i)   the frequency
(ii) the r.m.s. value
(iii) the instantaneous value after 1 ms.

(b) Calculate:- (i)   the total impedance
(ii) the power developed in the circuit
(iii) the r.m.s. current.

Fig. 79

277

ELECTRICAL AND ELECTRONIC PRINCIPLES II

INDEX